BIG IDEAS

from

Dr. Small

CREATING A COMFORT ZONE
FOR TEACHING MATHEMATICS

GRADES 4-8

D1605387

REVIEWERS

Iain Brodie, *Toronto District School Board*
Joan O'Keefe, *Toronto Catholic District School Board*
Susan Pitre, *Toronto District School Board*
Gladys Sterenberg, *University of Alberta*
Joyce Tonner, *Thames Valley District School Board*

NELSON PROFESSIONAL LEARNING

NELSON EDUCATION

Big Ideas from Dr. Small
Creating a Comfort Zone for
Teaching Mathematics Grades 4–8

Author
Marian Small

Director of Publishing
Kevin Martindale

General Manager, Mathematics, Science, and Technology
Lenore Brooks

General Manager, Professional Learning
Audrey Wearn

Publisher, Numeracy
Colin Garnham

Associate Publisher, Mathematics
Sandra McTavish

Executive Managing Editor, Development
Cheryl Turner

Managing Editor, Development
David Spiegel

Product Manager
Linda Krepinsky

Program Manager
Caron MacMenamin

Developmental Editor
Jackie Williams

Assistant Editor
Sarah Brown

Director, Content and Media Production
Linh Vu

Senior Content Production Manager
Sujata Singh

Content Production Editor
Shana Hayes

Copy Editor
Linda Jenkins

Proofreader
Cheryl Tiongson

Indexer
Andrew Little

Production Manager
Helen Jager-Locsin

Production Coordinator
Susan Ure

Design Director
Ken Phipps

Interior Design
Peter Papayanakis

Cover Design
Ken Phipps and Rocket Design

Cover Image
Veer.com

Asset Coordinators
Suzanne Peden
Renée Forde

Compositor
Nesbitt Graphics, Inc.

Printer
TC

Reviewers
The authors and publisher gratefully acknowledge the contributions of the following educators:
Iain Brodie, Toronto District School Board
Joan O'Keefe, Toronto Catholic District School Board
Susan Pitre, Toronto District School Board
Joyce Tonner, Thames Valley District School Board

Table of Contents

Chapter 2 Number and Operations 15

Chapter 3 Geometry   93

Shapes and Their Properties   93

BIG IDEAS FOR SHAPES AND THEIR PROPERTIES   93

Probability   197

Acknowledgments

We are fortunate in Canada to have ministries of education and teachers of mathematics who are open to progressive thinking about mathematics education. Because of this, Canadian teachers are willing to tackle learning more about mathematics in order to help their students.

I would like to thank those many committed educators for their interest in what I have shared with them over the years about how to make mathematics more meaningful to them and their students.

I would also like to thank Nelson Education Ltd. for being so supportive of my work and showing so much faith in me over the course of our many projects together. This book is due, in no small part, to the encouragement and efforts of a number of people at Nelson, particularly Audrey Wearn, director of professional learning, and my editor, Jackie Williams. Audrey is a consummate professional, savvy, easy to work with, and full of integrity. Jackie's keen eye and thoughtful approach ensures that what we produce is clear, coherent, and helpful to teachers. I would also like to acknowledge Caron Macmenamin, who worked with the editor, to create a lovely, accessible design and high quality DVD.

I would also like to thank the reviewers for their advice, helpful comments, and suggestions.

About the Author

Marian Small is the former Dean of Education at the University of New Brunswick. She has been a professor of mathematics education for over 30 years and is a regular speaker on K to 12 mathematics throughout Canada and the United States.

Over the last 20 years, she has been an author on seven student text series at both elementary and secondary levels in Canada, the United States, Australia, and the country of Bhutan, and a senior author on five of those series. She has served on the author team for the National Council of Teachers of Mathematics Navigation series, Pre–K to 2; for four years, she has been the NCTM representative on the Mathcounts question writing committee for middle school mathematics competitions throughout the United States. She is also a member of the editorial panel for the NCTM 2011 yearbook on motivation and disposition.

She has recently completed a text for university pre-service teachers: *Making Math Meaningful for Canadian Students: K–8;* a book on differentiated instruction in mathematics K to 8, *Good Questions: Great Ways to Differentiate Instruction*, published by Teachers College Press; and the companion book to this one: *Big Ideas from Dr. Small, Grades 4–8*.

She has led the research resulting in the creation of maps describing student mathematical development in each of the five mathematical strands for the K to 6 level and created the associated professional development program, PRIME. She has also developed materials and provided consultation for working with struggling learners and asking good questions.

Introduction

Issues in Teaching Mathematics

Teaching mathematics to Grades 4 to 8 students can be a challenge. Because the mathematics curriculum has become increasingly sophisticated over the last number of years, many of those teaching at these levels did not have the benefit of a grounding in some of the content they must teach. Even where the content is familiar, the approaches to teaching mathematics that the curriculums across the country require are a far cry from what many teachers have experienced themselves. In other words, teaching math, even to young students, can be complex and challenging.

The mathematics and teaching approaches have become more sophisticated over the years.

The Power in Being Comfortable with the Math We Teach

The question is, what can we do to make the challenge more manageable and productive? It might help to look at a metaphor. All of us know from the many facets of our lives how much more successful we are in accomplishing tasks when we are confident and comfortable with the situation. We also know that that air of confidence is usually sensed by those around us and often results in a level of comfort for them as well.

For example, when you drive visitors around your neighbourhood, you have an intuitive sense of what to do and your passengers are confident that you have the background knowledge that will guide you to your destination. However, when you take them to locations that are unfamiliar to you, they sense that you are tentative and reliant on "rules" (in this case maps) to which you repeatedly refer. So, unlike when you are in your comfort zone where your passengers are at ease, they again take their cues from you, except this time they are uneasy about how, and if, they will reach their destination.

Similarly, our students are our "passengers." Our job is to build their confidence and reassure them that they really can understand the new ideas with which they are dealing. This is accomplished by helping them create an internal map of how these new ideas connect to what they already know. This means we, as teachers, need that internal map, too, in order for our students to recognize our comfort and confidence.

Our job as teachers is to imbue our students with the confidence they need to understand the new ideas with which they are dealing.

What Teaching to Big Ideas Means

One of the gaps in many teachers' mathematical backgrounds is an internal map of the subject. They lack a fundamental understanding of how various mathematical topics interconnect, which topics are more important in the long term than others, and which aspects of those topics are most important. As they teach, many teachers often feel like they are going through a checklist, checking off whether students have learned each new discrete

Using a checklist of discrete outcomes/expectations to teach is in stark contrast to what we know from research about how students learn most effectively by making connections.

concept or skill listed in the curriculum. This is in stark contrast to what we know from research about how much more effective it is for students to learn when connections are explicitly made between new knowledge and ideas that students already know (Borko and Putnam, 1995; Schifter, Bastable, and Russell, 1997; Kennedy, 1997). Big ideas help teachers and students make these connections.

For example, instead of introducing decimals to students as a completely new number system, it is important for students to think of decimals as both an alternative representation to fractions and an extension of the whole number, base ten, place value system, content with which they are already familiar. This is the focus of the first big idea for Decimals on page 61 (Decimals are an alternative representation to fractions, but one that allows for modelling, comparisons, and calculations that are consistent with whole numbers, because decimals extend the pattern of the base ten place value system).

Another challenge for teachers is interpreting their curriculum. Big ideas can help here as well. Some curriculum outcomes/expectations are broader and more encompassing than others, but it is not obvious when you look at the curriculum. Using big ideas can help teachers assess the attention or emphasis required to achieve various outcomes/expectations. Big ideas also helps the teacher understand what some of the broader outcomes/expectations mean. For example, suppose the curricular goal is "solve problems requiring conversion from smaller to larger metric units." It may not be immediately obvious to the teacher that one of the main reasons why we convert measurements is related to our ability to use benchmark measurements to interpret and communicate measurements effectively (Big Idea 4 on page 129: Familiarity with known benchmark measurements can help you estimate and calculate other measurements). That is, conversion from a smaller unit to a larger is often about using the benchmark measurements with which most people are familiar. For example, it is easier for most of us to visualize what 3.1 m "looks like" than 310 cm or certainly 3100 mm because 3 m is a benchmark length; whereas 300 cm or 3000 mm is not. If a teacher understands this is the big idea behind the expectation/outcome, he or she can teach the concepts and skills involved in conversion much more meaningfully instead of a more superficial skills-based approach.

Another way that big ideas can help is to ensure teachers and students understand the overriding purpose behind a lesson or task. Teachers often use tasks that appear to be successful because they are very engaging to students. However, the reason why the students are doing the task is not clear and the teacher does not know what to focus on in his or her questioning throughout the task and at the conclusion of the task. Teaching through big ideas is about teachers looking critically at a task or lesson, asking themselves why they are teaching that lesson or task, and then making sure that purpose becomes clear to students so that the task or lesson can be more effective. For example, many teachers use tasks involving pattern block puzzles when teaching geometry (e.g., as shown at the top of the next page) without realizing that the reason these tasks are used is to focus student attention on the attributes of the shape they are using (Big Idea 3 for Shapes and Their Properties).

How shapes can be combined into other shapes helps students attend to the properties of the shapes (Big Idea 3 for Shapes and Their Properties on page 93), for example, which shapes have angles that can fit together to make a bigger shape and where the equal sides are.

Just as one would generally not teach a curriculum in its order of presentation, one does not teach big ideas in any specific order. It is simply that, in the course of teaching a curriculum in a meaningful and practical way, whether it is how a text resource does it, how a board or district has agreed to do it, or how a single teacher chooses to do it, the teacher keeps returning to the big ideas by choosing and shaping tasks and asking questions that keep bringing the big ideas to the forefront.

Despite increasing abstraction as students go up the grades, and shifts in focus from grade to grade, big ideas remain fundamentally simple and are the same from grade to grade. For example, the second Big Idea for Patterns on page 1, that the mathematical structure of a pattern can be represented in a variety of ways, applies whether students are creating simple shape patterns or more complex numerical patterns. Another example is, in any probability situation, it is important that students recognize that they can never be sure what happens next (Big Idea 3 for Probability), whether that student is doing a simple experiment in Grade 4 or 5 or a more complex simulation in Grade 7 or 8.

What is most important, though, is that the big ideas not be implicit, but explicit. Whether it is through the task itself, through teacher questioning

Big ideas remain fundamentally simple and are the same from grade to grade through the elementary school system.

Whether these students are using experimental probability or theoretical probability to predict what will happen next, they know that they cannot be certain of what will happen (Big Idea 3 for Probability on page 197).

and prompting, or through teacher talk, it is important that big ideas be said out loud. The more students hear an idea, the more likely it is that they will internalize it and be able to use it to support further learning. For example, within the Geometry strand, Big Idea 4 on page 93 states that many geometric properties and attributes are related to measurement. While students are learning about classification of triangles, a question such as "What measurements do you need to know in order to determine if a triangle is scalene?" would bring that big idea to the surface.

A teacher should also plan both instruction and assessment by considering big ideas. For example, in planning a unit of instruction for geometry for Grade 4, a teacher might consider which of the big ideas are relevant to mathematical content in that unit. The teacher would then plan instruction to ensure that each day he/she asks questions that elicit an explicit discussion of those big ideas. In planning assessment for the unit, the teacher would develop appropriate assessment tasks around the relevant big ideas. Too often our assessment focuses on the details and often the less important aspects of the mathematics when it is the big picture that really matters.

What is most important for us to assess is student knowledge of the enduring understandings and the big ideas (Wiggins and McTighe, 1999).

How This Resource Is Organized

This book is organized to focus on the big ideas in Grades 4 to 8 mathematics. Each chapter focuses on one of the five mathematical strands and each chapter is then divided into familiar topics. For each topic in each strand, a set of big ideas that underlie the curriculum is articulated. For example,

Chapter 2   Number and Operations
    Topics: Numbers Beyond 1000
              Number Theory
              Whole Number Operations
              Fractions
              Decimals
              Proportional Thinking: Rate, Ratio, and Percent
              Integers

The main body of text helps the teacher better understand the mathematics that they need to know in order for their students to be successful in that strand.

Teaching Ideas are tasks and questions related to the mathematics being discussed that the teacher can use to bring the big ideas to the surface.

A special feature of the book is the Teaching Ideas in the margins. These include tasks and strategies related to the mathematics being discussed in the main body beside them, along with questions that the teacher can ask to bring the big ideas to the surface. These Teaching Ideas are intended as model tasks with model questions to elicit big ideas and many are adaptable to similar situations at another grade level that might be of more interest to a particular reader.

The example at the top of the next page from page 172 will help you understand how the Teaching Ideas are structured. Notice that specific grade levels have not been assigned to the Teaching Ideas. This is because specific content is taught at different grade levels across the country. Because the relevant content is discussed right beside the Teaching Idea, the reader should have no difficulty recognizing how the Teaching Idea relates to the content they teach and should be able to judge its appropriateness for their grade level.

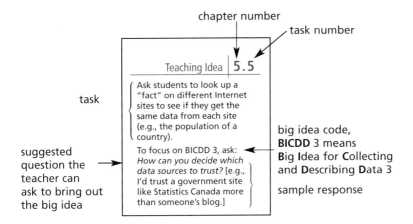

chapter number

task number

Teaching Idea | **5.5**

task

Ask students to look up a "fact" on different Internet sites to see if they get the same data from each site (e.g., the population of a country).

suggested question the teacher can ask to bring out the big idea

To focus on BICDD 3, ask: *How can you decide which data sources to trust?* [e.g., I'd trust a government site like Statistics Canada more than someone's blog.]

big idea code, **BICDD** 3 means **B**ig **I**dea for **C**ollecting and **D**escribing **D**ata 3

sample response

Beside the Teaching Idea in the main body of the text, second-hand data as a source is discussed.

Other Features of This Resource

In many, but not all, of the chapters, there are lists of principles or important points that will help the teacher better organize instruction. These principles and important points are generally not as broad as the big ideas, but they are important concepts to explicitly address in instruction, points that may not be obvious to all teachers.

For example, the list of Principles for Comparing Fractions on page 48 refers to several specific ideas around comparing fractions. Two examples are, if two fractions have the same denominator, the one with the greater numerator is greater, and if two fractions have the same numerator, the one with the greater denominator is less. These ideas are not as broad as the one that is identified as the big idea, that renaming fractions is often the key to comparing them or computing with them (Big Idea 3 on page 42), but they will help the teacher teach more effectively to address that big idea. Similarly, the list of Important Points About Measures of Central Tendency and Data Spread in Chapter 5 (on page 178) helps the teacher know where to focus instruction in order to address Big Idea 4 (page 167), that a large set of data can be usefully described by a single summary statistic or a combination of statistics. For example, students could explore whether the mean, median, or mode alone or the range and mean, range and median, or range and mode gives the better summary of a set of data.

A glossary is provided at the end of the book should the reader encounter a mathematical term with which he or she is not familiar. It will also serve as a handy reference for ongoing mathematics teaching.

Although the focus of this resource is on helping teachers understand the mathematics and teach through big ideas, the resource also reflects many other important instructional considerations, including

- the use of effective, and multiple, representations for mathematical concepts;
- the role of manipulatives in making math meaningful to students;
- the value of multiple solutions;
- the value and the potential range of personal strategies and invented algorithms; and
- the importance of teaching conceptually.

Chapter 1

Patterns and Algebra

Patterns

Students begin working with patterns in early elementary school. Although a focus on algebra replaces a focus on patterns at the secondary level, some attention to patterns remains throughout the grades.

BIG IDEAS FOR PATTERNS

1. Patterns represent identified regularities. There is always an element of repetition, whether the same items repeat, or whether a "transformation," for example, adding 1, repeats.

2. The mathematical structure of a pattern can be represented in a variety of ways.

3. Some ways of displaying data highlight patterns.

4. Many geometric attributes, measurements, and calculations involving numbers are simplified by using patterns.

Types of Patterns

Complex Repeating Patterns

Repeating patterns that involve more than one attribute are appropriate for study by Grades 4 to 8 students. For example, the pattern below involves both shape and colour. Notice that the colour follows an AAB pattern (blue, blue, yellow), but the shape follows an AB pattern (circle, square).

A complex multi-attribute pattern using shape and colour—
shape is AB, while colour is AAB.

The fact that the two attributes follow different pattern rules is what makes this type of pattern more appropriate for Grades 4 to 8 students than for younger students. Younger, or struggling, students are more likely to be able to deal with two-attribute patterns where both attributes follow the same pattern rule. For example,

A simple multi-attribute pattern using shape and colour.
Both shape and colour follow an AB pattern.

Teaching Idea | **1.1**

Ask students to create a shape pattern for which the number of shapes in each term forms an ABA pattern, but the terms themselves form an AB pattern, such as:

1, 2, 1, 1, 2, 1, ...
circle, square, circle, square, ...

Bring out BIP 1 by asking: *What is it that makes it a pattern? If I changed this shape, would it still be a pattern?*

Teaching Idea | 1.2

Provide several, but not all, terms of a pattern except the first one, such as those shown below. Tell students that, in each case, the numbers increase or decrease by a constant amount. Students can then work forward and backwards to determine the missing terms.

_, _, 9 , _, _,15,
[5, 7, 9 , 11, 13,15, 17]
or

_, 36, _, 28, _, 20
[40, 36, 32, 28, 24, 20]

After students are done, ask: *How did you figure out what the decrease was each time? Why did you need to add and subtract?*

Notice that, as is suggested by BIP 4, pattern work provides a context for other mathematical topics, such as number.

Teaching Idea | 1.3

There are many interesting ways to show the pattern 1, 4, 9, ..., including

• a pattern of squares of increasing size like this:

• a step pattern:

Show the two patterns. Point out to students that the same pattern is represented in different ways (BIP 2). One representation emphasizes the squared nature of the numbers while the other emphasizes that it is the sum of increasing odd numbers. Ask students to find another representation. [e.g., the triangle pattern on the right]

Bring out BIP 3 by asking: *Which pattern makes you think of the numbers in terms of multiplication? addition? Why?* [squares, I see 1 × 1, 2 × 2, and 3 × 3; triangles and steps, I see +3, +5]

Students in Grades 4 to 8 are also more able to deal with patterns that go in more than one direction, like the one shown below. This type of pattern is commonly observed in our environment.

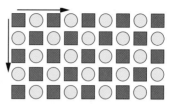

A two-dimensional pattern using shape and colour

Increasing and Decreasing Patterns

Some increasing patterns are quite familiar to students, particularly the number pattern 1, 2, 3, 4, ... As children develop mathematically, they experience other increasing (and decreasing) patterns. These patterns may be

• arithmetic sequences, where each number is a fixed amount greater or less than the preceding one
  3, 5, 7, 9, ... (fixed increase of 2)
  12, 10, 8, 6, ... (fixed decrease of 2)
• geometric sequences, where each number is a fixed multiple of the preceding one
  2, 4, 8, 16, ... (doubling pattern)
  100, 20, 4, ... (dividing by 5 or multiplying by $\frac{1}{5}$ pattern)
• other number sequences where the increase is not constant
  3, 4, 6, 9, 13, ... (increase by 1 more each time)
  20, 18, 14, 8, ... (decrease by 2 more each time)

Students are generally able to recognize and extend arithmetic sequences before they are comfortable with the other types of increasing and decreasing sequences, because arithmetic sequences are based on simply adding or subtracting the same amount each time. There can also be increasing shape patterns, such as the patterns below.

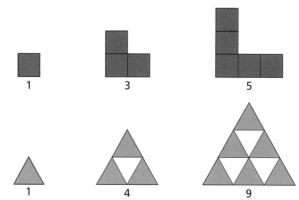

Some shape patterns can also be viewed as number patterns.

These increasing shape patterns can often be represented and described as increasing number patterns: 1, 3, 5, ... and 1, 4, 9, ... Notice that each increasing or decreasing pattern involves repetition of how the increase or decrease occurs, which is what makes it a pattern.

Be aware that students tend to struggle more with decreasing patterns than with increasing patterns. Perhaps it is our job to deal more with these types of patterns than we currently do.

Recursive Patterns

A recursive pattern is one where each element in the pattern is defined based on a previous item or items. Some of the growing patterns described in the previous section can be defined recursively. For example, for the sequence 2, 4, 6, 8, ..., each term is defined as 2 greater than the preceding term. However, some recursive patterns are more complex. An example of a recursive pattern that is not of that type is the pattern 2, 3, 6, 18, 108, ..., in which each term is defined as the product of the two preceding numbers.

Some particularly famous recursive sequences include

- the triangular numbers—beginning at 1, each term is created by adding 1 greater than was added previously: 1, 3, 6, 10, 15, ... (Start at 1, add 2, then 3, then 4, then 5, ...)
- the Fibonacci sequence—beginning with 1 and 1, add two terms to get the next term: 1, 1, 2, 3, 5, 8, 13, 21, 34, ... (Start with 1 and 1; add the first two terms to get the 3rd term; then add the 2nd and 3rd terms to get the 4th term, then add the 3rd and 4th terms to get the 5th term, ...)

Extending and Describing Patterns

Extending Patterns

Once students have identified a pattern, you normally ask them to either extend it or describe it so that you can see that they understand the pattern. It is often easier for students to demonstrate their recognition of the pattern by extending it rather than describing it.

Sometimes when you ask students to extend a pattern, there is discrepancy between what you expect and what students understand or perceive. For example,

- Students legitimately view the pattern in a different way than you expect because no pattern rule is provided or not enough elements have been included. For example, a teacher asks a student to extend the pattern 2, 3, 5, 8, ... The teacher is expecting the student to say 13, 21, ... (where each term of the pattern is the sum of the two preceding terms), but there are many other legitimate ways to extend the pattern, such as 2, 3, 5, 8, 2, 3, 5, 8, ... (repeating a four-term core) 2, 3, 5, 8, 12, 13, 15, 18, 22, 23, 25, 28, ... (adding 10 to each of a group of terms) 2, 3, 5, 8, 8, 5, 3, 2, 2, 3, 5, 8, 8, 5, 3, 2, ... (repeating four terms forward and backwards) 2, 3, 5, 8, 12, 17, ... (adding 1 first, and then adding 1 more each time)
- Students seem to find it easier to extend a pattern that ends with a full repetition of the core, rather than in the middle of the core. For example, students might have difficulty extending the pattern *circle, circle, square, circle, circle, square, circle, circle, square, circle, ...* and will find it much easier if full repetitions of the core have occurred, as in *circle, circle, square, circle, circle, square, circle, circle, square, ...*.

Describing Patterns

Teachers often ask students to describe a pattern to ensure that they understand it. Ideally, a pattern's description is a *pattern rule,* whether described verbally, pictorially, or symbolically. The rule is an unambiguous description of the pattern.

Teaching Idea | 1.4

Have students act out the "handshake" problem. Students, in groups of 2, 3, 4, and 5, each shake hands with each other student once and complete a chart like the one shown below:

STUDENTS	HANDSHAKES
2	
3	
4	
5	

The number of handshakes forms the pattern of triangular numbers [1, 3, 6, 10]. This activity shows a different physical representation of the numerical pattern (BIP 2).

When students have completed the chart, ask: *Predict the number of handshakes for 6 students and 9 students.* [15; 45] *How did you use the pattern to figure that out?* [I continued the pattern +2, +3, +4, +5, ...]

Teaching Idea | 1.5

Ask students to extend a pattern that begins with two numbers, like 5, 10, ... , in four different ways. [e.g., 5, 10, 15, ... ; 5, 10, 5, 10, 5, 10, ... ; 5, 10, 20, ... ; 5, 10, 16, 23, 31, ...] Emphasize that a pattern must involve some repetition or regularity (BIP 1).

Once students have completed and described their patterns, ask: *If we had started with three numbers, like 5, 10, 15 ..., would there only be one possible pattern? How do you know?* [no; e.g., 5, 10, 15, 20, ...; 5, 10, 15, 5, 10, 15, ...; 5, 10, 15, 25, 30, 35, 45, 50, 55, ...; 5, 10, 15, 21, 27, 34, 41, ...]

It is important to stress that, without a pattern rule, you cannot extend a pattern with confidence, since there is an unlimited number of possible patterns, even for a pattern for which you know the first three terms (this is a common misconception). This is demonstrated in **Teaching Idea 1.5**. A pattern rule describes one possible pattern only. If more than one pattern is possible, then it is not a pattern rule.

Frequently, students focus on one aspect of the description, but forget another important part of it. For example, if a student describes the pattern 4, 7, 10, 13, ... as "an add 3 pattern" without indicating that it starts at 4, the pattern rule is incomplete. Clearly, this is not a problem if the first part of the pattern is shown, so it may seem superfluous to say that it starts at 4. Formally, students should learn that a pattern rule must describe each and every element of the pattern (including the first element).

Students might find it easier to describe a pattern by comparing it to another pattern. For example, "the pattern 3, 6, 9, 12, 15, ... is like 2, 4, 6, 8, ... but starts at 3 instead of 2 and goes up by 3 instead of by 2."

Translating and Representing Patterns

Translating Patterns

To determine whether a student is focused on the mathematical structure of a pattern, it is natural to ask the student to translate a pattern into a different, but equivalent, form. For example, a student might be shown the pattern below using brown, blue, and white squares and triangles, and asked to show the same kind of pattern using red, yellow, and green circles and squares. The pattern is ABB for shape and ABC for colour. There are many possibilities, as shown below.

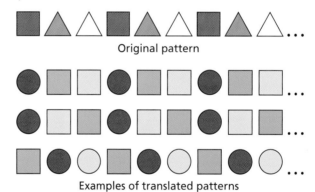

Original pattern

Examples of translated patterns

It is much easier to translate repeating patterns than increasing, decreasing, or recursive patterns, since in the latter cases the amount of increase or decrease or the first number might actually be viewed as part of the structure of the pattern. However, students can systematically alter the elements of a pattern rule to see how a new pattern relates to the original pattern, even though it might not be a direct translation. This will be built on later as students begin to work with linear, quadratic, and exponential functions.

Representing Patterns

For increasing and decreasing numerical patterns, using geometric representations can help highlight the pattern. For example, the numerical pattern 4, 7, 10, 13, ... can be represented geometrically as shown at the top of the next page. This brings out clearly that the pattern increases by 3 each time.

Ask students to write the pattern rule for the pattern 3, 5, 7, 9, …, and then apply BIP 2 by representing the pattern using shapes. Have them "translate" the numerical pattern by altering either the first term or the increase amount and then representing the new pattern with the same shapes. They can compare the two shape patterns to see the effect of the translation. You might ask: *How does the difference in the two shape patterns show the difference in the two number patterns?*

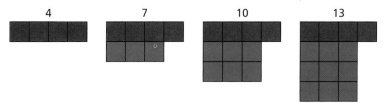

This visual representation shows the pattern:
4 is the start number and 3 is added each time.

The same visual representation coloured differently can highlight the relationship between the term value (number of squares) and term number (or position). This clearly shows that, to find the number of squares in a term, you can triple the term number and add 1.

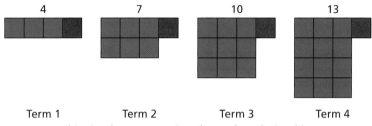

| Term 1 | Term 2 | Term 3 | Term 4 |

This visual representation shows the relationship:
term value = 3 × term number + 1

Creating Patterns

Students should be encouraged to create patterns of various types using their own pattern rules. The rules can describe the repetition, the growth or decrease, the value of a particular term, etc. By exposure to many types of patterns and patterns of various structures based on many different attributes, students' own patterns will be richer.

Using Patterns to Develop Mathematical Concepts

Much of the mathematics that students study is built on pattern—students often learn how to calculate using patterns; patterns are inherent in measurement (such as metric unit prefixes); and patterns are used in geometry (e.g., to relate the number of diagonals in a regular polygon to its number of sides). It is important to point out the pervasiveness of pattern in the study of mathematics as these ideas are being developed. A few examples of how patterns are used in the number strand are shown here.

Teaching Idea | 1.7

Ask students to choose a criterion from this list to create a pattern (BIP 1):

- Use three different shapes to create a decreasing pattern.
- Create an increasing pattern with a tenth term of 52.
- Create a pattern that grows, but not by the same amount each time.
- Create a decreasing pattern with a fifth term of 32.

When students are done, ask: *Could you have created a different pattern? How do you know?*

Patterns Used in Number

EXAMPLE	COMPUTATIONAL RELATIONSHIP
31 + 5 = 36 41 + 5 = 46 51 + 5 = 56 61 + 5 = ☐	As students complete each answer, the patterns inherent in the place value system should become apparent. When adding the same amount to a number that is 10 greater, the sum is 10 greater; the tens digit changes, but the ones digit does not. Thus, the sum of 61 + 5 must be 66.
32 − 18 = 14 42 − 28 = 14 52 − 38 = 14 62 − 48 = ☐	As students complete the subtractions, they should notice that when both the minuend and the subtrahend increase by the same amount, the difference is the same. Thus, the difference for 62 − 48 must be 14.

(continued)

Patterns Used in Number (continued)

EXAMPLE	COMPUTATIONAL RELATIONSHIP
$4 \times 3 = 12$ $4 \times 30 = 120$ $4 \times 300 = 1200$ $4 \times 3000 = \square$	As students determine each answer, they should observe that each time an additional 0 appears at the end of the second factor, there is an additional 0 at the end of the product. Thus, the product of 4×3000 must be 12 000. Although this pattern does not explain why this is so, it is a good visual model to start off the conversation about multiplying by powers of 10.
$3 \times 2 + 3 \times 3 = 3 \times 5$ $3 \times 2 + 3 \times 4 = 3 \times 6$ $3 \times 2 + 3 \times 5 = 3 \times 7$ $3 \times 2 + 3 \times 6 = 3 \times \square$	As students determine each answer, they should observe that the second factor on the right is the sum of the two red numbers on the left side of the equation. Thus, the answer to $3 \times 2 + 3 \times 6$ must be 3×8. Again, this is not an explanation, but it piques the curiosity of students so that the concept can be explored and reasoned through.
$4 \times 2 = 8$ $4 \times 1 = 4$ $4 \times 0 = 0$ $4 \times (-1) = \square$	At the middle school level, patterns are often used to explain operations with integers. Based on the mathematical pattern inherent in the first three products, students should see that $4 \times (-1)$ must be -4.

Patterns Relating Fractions to Decimals

There are an extraordinary number of patterns that can be uncovered in working with fractions. One very interesting pattern emerges when students explore the decimal expressions for what are called the unit fractions—the fractions with numerator 1.

A bar over digits in a decimal indicates that the digits repeat. For example, the 6 in $0.1\overline{6}$ repeats ($0.166\,66 \ldots$), and the sequence 142 857 in $0.\overline{142\,857}$ repeats ($0.142\,857\,142\,857\,142\,857 \ldots$).

Students might observe patterns in the decimal equivalents for unit fractions up to $\frac{1}{20}$ and for fractions with denominators of 9, 99, and 999.

Patterns in Fraction 9ths, 99ths, and 999ths

9THS	99THS	999THS
$\frac{1}{9} = 0.\overline{1}$	$\frac{1}{99} = 0.\overline{01}$	$\frac{1}{999} = 0.\overline{001}$
$\frac{2}{9} = 0.\overline{2}$	$\frac{2}{99} = 0.\overline{02}$	$\frac{2}{999} = 0.\overline{002}$
$\frac{3}{9} = 0.\overline{3}$	$\frac{3}{99} = 0.\overline{03}$	$\frac{3}{999} = 0.\overline{003}$

Teaching Idea | 1.8

Students can explore the patterns in the decimal equivalents for $\frac{1}{11}, \frac{2}{11}, \frac{3}{11}, \ldots, \frac{9}{11}$ [0.09 …, 0.18 …, 0.27 …, …, 0.81, …] and for $\frac{1}{101}, \frac{2}{101}, \frac{3}{101}, \ldots, \frac{9}{101}$ [0.0099 …, 0.0198 …, 0.0297 …, …, 0.0891, …] using a calculator (BIP 4).

When students have completed each pattern, ask: *How are the two patterns related?* [both involve digits that add to multiples of 9; both go up in one place value and down in another] *How could you use the patterns to predict the decimals for $\frac{1}{1001}, \frac{2}{1001}, \frac{3}{1001}, \ldots$?* [I predict 0.000 999, 0.001 998, …, since instead of one and two zeroes after the decimal there would be three zeroes, and instead of going 09, 18, 27, … or 0099, 0198, 0297, …, it makes sense it would be 000 999, 001 998, 002 997, …]

Patterns in Unit Fractions

FRACTIONS AND DECIMALS	PATTERNS
$\frac{1}{3} = 0.\overline{3}$ $\frac{1}{4} = 0.25$ $\frac{1}{5} = 0.2$ $\frac{1}{6} = 0.1\overline{6}$ $\frac{1}{7} = 0.\overline{142857}$ $\frac{1}{8} = 0.125$ $\frac{1}{9} = 0.\overline{1}$ $\frac{1}{10} = 0.1$ $\frac{1}{11} = 0.\overline{09}$ $\frac{1}{12} = 0.08\overline{3}$ $\frac{1}{13} = 0.\overline{076\,923}$ $\frac{1}{14} = 0.0\overline{71\,428\,5}$ $\frac{1}{15} = 0.0\overline{6}$ $\frac{1}{16} = 0.0625$ $\frac{1}{17} = 0.\overline{058\,823\,529\,411\,764\,7}$ $\frac{1}{18} = 0.0\overline{5}$ $\frac{1}{19} = 0.\overline{052\,631\,578\,947\,368\,421}$ $\frac{1}{20} = 0.05$	• Reading down the list of fractions, the first digit after the decimal point either stays the same or decreases. (This is because the fraction equivalents are decreasing in value, so the decimals decrease as well.) • For any unit fraction, the number of digits appearing in the repeating cycle of the decimal is the same as the number of digits appearing in the repeating cycle for the unit fraction with a denominator twice as great. For example, there is 1 repeating digit in the decimal for $\frac{1}{6}$ ($0.166\,666 \ldots$), 1 repeating digit in the decimal for $\frac{1}{12}$ ($0.083\,333 \ldots$), 6 repeating digits in the decimal for $\frac{1}{7}$ ($0.142\,857\,142\,857 \ldots$), and 6 repeating digits in the decimal for $\frac{1}{14}$ ($0.071\,428\,571\,428\,5 \ldots$). • Students might observe that the number of digits that repeat for all the prime denominators except 2 and 5 is a factor of 1 less than the denominator. For example, there are 6 repeating digits for $\frac{1}{7}$ (6 is a factor of $7 - 1$), and 6 repeating digits for $\frac{1}{13}$ (6 is a factor of $13 - 1$).

Algebra

Work on patterns leads naturally to algebraic thinking. And, just as pattern work relates other mathematical strands, so does algebra. For example, the development of formulas, which are algebraic equations, is an important part of the measurement strand.

BIG IDEAS FOR ALGEBRA

1. Algebra is a way to represent and explain mathematical relationships and to describe and analyze change.

2. Using variables is a way to efficiently and generally describe relationships that can also be described using words.

Each teaching idea in this section of the chapter will indicate which Big Idea(s) for Algebra (BIA) can be emphasized.

What Is Algebra?

Most people think that algebra is math with letters. However, mathematicians view algebra as a system that allows one to

- represent mathematical relationships
- explain relationships among quantities
- analyze change

For example, when you write the formula $P = 4s$ for the perimeter of a square, you are representing a mathematical relationship. By exploring the formula and noting that, for example, as s increases by 1, P increases by 4, you are exploring the relationship between the side length and the perimeter. You are simultaneously exploring the notion of change.

Variables in Algebraic Thinking

It is not the use of letters to represent quantities that defines algebra; in fact, students are also thinking algebraically when they solve open number sentences like $4 + 3 = \square$ or $7 - \square = 3$, which use boxes (or open frames) instead of letters. When letters or open frames are used in mathematics, you say you are using *variables*.

The convention, when using letters, is to use lowercase letters to represent variables. Mathematicians generally use the letter n when representing whole numbers; however, other popular letters are x, y, and z. The progression is usually from the use of the open frame ($\square$) to letters. Variables, however, are used for different purposes, and this creates some confusion among students.

Using Variables to Represent Unknown Values

In the equation $x + 5 = 8$, or $\square + 5 = 8$, the variable represents a specific value yet to be determined. Sometimes there is only one possible value for the unknown, sometimes there are no possible values, and sometimes there are multiple values or an infinite number of values.

Teaching Idea | **1.9**

Ask students to compare the number of whole number values greater than 0 that make each true.

$\square + 5 = 10$ [1 value: 5]

$\square + 5 < 10$ [5 values: 0, 1, 2, 3, 4]

$\square + 5 = 5 + \square$ [any value: 0, 1, 2, 3, ...]

Discuss the idea that a variable can play different roles in an equation by asking: *Can you make another sentence that is true for only one value? for several values? for any value?*

Talk about how equations using variables represent relationships between numbers (BIA 2) by asking: *What number relationships does the equation* $6 \times 4 = 12 \times \square$ *show?* [12 is twice 6, so $\square$ is half of 4]

Using Variables to Generalize

Mathematicians are always looking for succinct, symbolic ways to describe multiple situations. For example, young students learn that "the order in which you add numbers doesn't matter" (the commutative property of addition). Although the words make it clear, mathematicians choose to describe this with symbols. A typical format would be $a + b = b + a$. This is a way of saying that no matter what numbers you use to replace a and b, if you add them in one order, you get the same result as adding in the other order. Some other examples of using variables to generalize relationships are listed below.

Common Algebraic Generalizations

ALGEBRAIC EXPRESSION	GENERALIZATION
$ab = ba$	You can multiply numbers in any order without changing the product (the commutative property of multiplication).
$2a + 2b = 2(a + b)$	The effect of doubling one number and adding it to double another number is the same as if you add the numbers first, and then double the sum (the distributive property of multiplication over addition).
$(n + 1)(n - 1) = n^2 - 1$	To calculate the product of the two numbers on either side of a number, you can square the middle number and subtract 1. For example, $11 \times 9 = 10^2 - 1$. This is very handy for some calculations; for example, you can quickly calculate 301×299 as $300^2 - 1 = 90\,000 - 1 = 89\,999$.
$\dfrac{a}{b} + \dfrac{c}{b} = \dfrac{(a + c)}{b}$	To add two fractions with the same denominator (in this case, b), you can add the numerators to determine the new numerator, and use the same denominator. This use of variables to generalize also describes a general computation process.
$4n$	To symbolize an arbitrary multiple of 4, a mathematician uses the expression $4n$, since $4n$ is the result of multiplying a whole number n by 4.

Simplifying Expressions

Students can represent algebraic expressions concretely using algebra tiles. If, for example, a yellow rectangle is used to represent n, then students soon realize that $2n$ can be represented by two yellow rectangles. Usually, one colour is used to represent positive amounts (for example, one yellow rectangle for n and one yellow square for 1), and another colour is used to represent negative amounts (for example, one white rectangle for $-n$ and one white square for -1).

Representing two algebraic expressions concretely using algebra tiles

The zero principle that is used to help students work with positive and negative integers is also applied when operating with algebraic expressions. So, for example, $2n - 3n = -n$ can be modelled as shown on the next page.

$2n - 3n = ?$

2n

To subtract 3n (the subtrahend), add tiles with a value of 0, using the zero property:

Take away 3n (3 yellow tiles):

$2n - 3n = -n$

To be able to subtract 3n from 2n,

- two tiles with a total value of 0 are added to 2n using the zero property,
- 3n is then subtracted,
- leaving $-1n$ or $-n$ as the difference.

The zero property was used because 3n could not be taken away from 2n. Adding one yellow rectangle ($+n$) and one white rectangle ($-n$) to the minuend did not change the value of the minuend value but it did allow for three yellow tiles, 3n, to be taken away.

Note that this model uses the takeaway meaning of subtraction (i.e., $2n - 3n$ means taking away 3n from 2n).

The missing addend meaning of subtraction could have been used instead (i.e., $2n - 3n$ means "How much do you have to add to 3n to get 2n? or 3n + ? = 2n). To do this, you would model 3n as three yellow tiles and then add one white tile ($-n$) to the model to make it 2n. That means $3n + (-n) = 2n$, so $2n - 3n = -n$.

Describing Relationships and Functions

A formula is a special algebraic equation that shows a relationship between two or more different quantities. It might be a formula

- relating speed (or rate), time, and distance ($D = rt$)
- describing the volume of a rectangular prism ($V = l \times w \times h$)
- relating the value of a term in a pattern to its position in the pattern; for example, in the pattern 2, 4, 6, 8, 10, 12, ..., each term's value is double its position in the pattern: $t = 2p$

One of the earliest student introductions to this notion of relationship, or *function*, involves the concepts of input and output. A function is a relationship that leads to a particular output for a particular input. For example, if the function is "double the number," the input is the number to be doubled and the output is the double. Often these inputs and outputs are shown in a table to make it easier for students to infer relationships.

Using a Table

A table (sometimes informally referred to as a t-chart because of its internal "T" shape) is a graphic organizer that allows students to describe a relationship or function by listing the inputs in one column and the corresponding outputs in the other. The table below shows the relationship "subtract 1 and double," as you subtract 1 from the number in the Input column, and then double the result to get the number in the Output column. In high school, this would be described as $f(x) = 2(x - 1)$, with $f(x)$ being function notation.

Teaching Idea | **1.10**

Let students choose a formula to work with or suggest a formula, such as $V = l \times w \times h$. Have them systemically change the value of one variable to see the effect on another. For example, they could change the value of h in $V = l \times w \times h$ from 1 to 2, then to 3, then to 4, ..., keeping the width and length the same, to see what happens to the volume.

To bring out BIA 1, ask: *How does the volume change when you double only the height?* [it doubles] *Why does that make sense?* [If a rectangular prism is twice as high as another, but the other dimensions are the same, it should be twice as big. I notice that if you double one side of the equation, you have to double the other side: $V = l \times w \times h \rightarrow 2 \times V = l \times w \times 2 \times h$.]

T-Chart Showing a Relationship

INPUT	OUTPUT
3	4
4	6
5	8
6	10

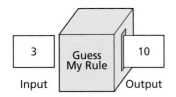

"It could be add 7, or multiply by 3 and add 1."

A table is an appropriate way to show the numbers in a pattern, with the input being the position in the pattern and the output the value of the number in the pattern. For example, the pattern 4, 7, 10, 13, 16, ... could be modelled in a table as shown below. To get each output value, each input value is multiplied by 3, and then 1 is added. Algebraically, this could be described as $o = 3i + 1$.

T-Chart for the Pattern 4, 7, 10, 13, 16, . . .		
	INPUT	**OUTPUT**
The position of the number in the pattern	1	4
	2	7
	3	10
	4	13

The value of the number in the pattern

In the table above, many students will see the pattern in the output column (4, 7, 10, 13, ...), but may not be able to figure out the relationship between the Input and Output columns. Determining this relationship can be difficult, but it is what will enable students to make predictions about the value of a number in a pattern based on its position.

Using a table to show how the value of the terms of a pattern relates to the position of the terms highlights the difference between a recursive pattern and relationships. When students see the intervals of 3 in the Output column above, they are thinking recursively. When students recognize that you can take the input number, triple it, and add 1 to get the output number, then students are thinking about a relationship. The ability to determine these relationships typically comes later than an understanding of recursive patterns in students' mathematical development, although students should be encouraged to read across the table, and not just up and down. Just as was the case in extending patterns, students can only be sure of the relationship rule if they are provided with enough specific information to rule out other possibilities.

Solving Equations

Students might be given the solution to an equation (e.g., $n = 10$), and then asked to create a variety of algebraic equations with that solution [e.g., $2n = 20$, $n + 7 = 17$] to apply BIA 2.

When students are done, ask: *How did you create your equation? How do you know that your equation is equivalent to the equation $n = 10$?*

Early in their mathematical careers, students solve open number sentences. Later, you refer to these as equations, but the idea is the same. An equation is a mathematical sentence with an equals sign. For some students, the equality sign poses a difficulty. Although they are comfortable with, for example, the sentence $4 + 5 = \square$, they interpret the equality sign to mean "find the answer." Therefore, when students see the sentence $\square - 4 = 5$, they may not be sure what to do, because they think the answer is already there. Similarly, students might solve $4 + \square = 5$ by adding 4 and 5 to "get the answer." The notion of an equation as an expression of balance is not apparent to them. This long-standing problem is exacerbated by the fact that many calculators require the = key to be pressed to get an answer, so students are reinforced in interpreting the = sign as synonymous with "get the answer."

It is important for students to recognize that the equality sign should be viewed as a way to say that the same number has two different names, one on either side of the equals sign.

Solving an Equation Using Manipulatives

When a number sentence is simple enough, students can call on a fact they already know or model it with manipulatives, such as counters, to determine the missing value. For example,

- To solve $4 \times 5 = \square$, students model 4 sets of 5 to calculate the 20.
- To solve $4 \times \square = 24$, students create 4 groups and keep putting out counters until 24 counters are placed in 4 equal-sized groups. Students then observe that each group includes 6 items.
- To solve $3 \times \square + 6 = 21$, students distribute 21 counters so that an equal number are in each of three circles and 6 are outside the circles. The solution is the number of counters in one circle [5], so $3 \times 5 + 6 = 21$, as shown below.

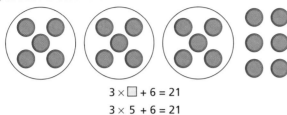

$$3 \times \square + 6 = 21$$
$$3 \times 5 + 6 = 21$$

Solving an Equation by Working Backwards

Students can think about working backwards to solve an equation like $4 \times \square = 24$. What the equation says is that a number is multiplied by 4 to get 24. The "reverse" of multiplying by 4 is dividing by 4. So students can work backwards and divide 24 by 4 to get 6.

Students might solve a more complex equation, such as $5 \times \square + 2 = 27$, by thinking "I multiply a number by 5, and then add 2 to get 27. So, if I work backwards and subtract 2 ($27 - 2 = 25$), and then divide by 5 ($25 \div 5 = 5$), I will end up with the solution, 5."

A more involved example of this strategy is found in **Teaching Idea 1.13**. After students apply the steps, they can follow the steps in reverse to see how the trick works.

Some equations, such as $2n + 3 = n + 5$, do not lend themselves to working backwards. Other strategies that students can use to solve equations like this are guessing and checking, or balancing.

Solving an Equation by Guessing and Checking

To solve $2n + 3 = n + 5$ ($n = 2$), a student might think the following.

Solving $2n + 3 = n + 5$		
GUESS	**REASONING**	**CHECK**
First guess: 10	I'll start with 10 because it's an easy number to work with mentally.	$2 \times 10 + 3 = 23$ $10 + 5 = 15$
Second guess: 5	I'll try 5, a number less than 10, because, if n is too great, doubling it and adding 3 (the left side of the equation) will make it a lot more than just adding 5 to it (the right side of the equation).	$2 \times 5 + 3 = 13$ $5 + 5 = 10$
Third guess: 4	I'll try 4 because I notice that 13 and 10, the numbers I got when I used 5, were closer together than when I used 10.	$2 \times 4 + 3 = 11$ $4 + 5 = 9$
Fourth guess: 3	I'll try 3 to get numbers that are closer together.	$2 \times 3 + 3 = 9$ $3 + 5 = 8$
Final guess: 2	I'll try 2 to get numbers that are closer together.	$2 \times 2 + 3 = 7$ $2 + 5 = 7$

Teaching Idea | **1.13**

Students can try this number trick, and then do it in reverse to see how the trick works.

- Select any number. Double it. Subtract 3. Double again. Add 2. Divide by 4. What number did you start and end with? [e.g., started with 5 and ended with 4]
- Now try the trick in reverse. Use your end number and work backwards. Predict the number you will get and then check your prediction. [5]

To focus on BIA 1, ask: *What did you do?* [I predicted 5. I multiplied 4 by 4, subtracted 2, took half, added 3, and then took half and got 5.] *Why does this happen when you do the trick in reverse?* [Addition and subtraction are inverse operations, as are multiplication and division.]

Solving an Equation by Maintaining a Balance

Students should start out using manipulatives such as a balance scale with paper bags representing the unknown number of cubes. For example,

MODELLING $2n + 3 = n + 5$	SOLVING BY MAINTAINING A BALANCE ($n = 2$)
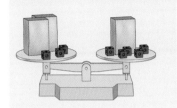	To model $2n + 3 = n + 5$, put 2 open, empty paper bags ($2n$) and 3 cubes on one side of a balance, and 1 open, empty paper bag (n) and 5 cubes on the other side. Then add cubes to the bags, ensuring there are the same number of cubes in each bag. Increase the number of cubes in the bags until the sides of the scale balance. The number of cubes in each bag is the value of the unknown, $n = 2$. $$2n + 3 = n + 5$$ $$2(2) + 3 = 2 + 5$$ $$7 = 7$$

On a symbolic level, the equation $2n + 3 = n + 5$ can still be considered a balance. If either side of the equation is changed, the other side must be changed the same way to maintain a balance. Therefore, if 3 is subtracted from both sides of the equation, $2n + 3 - 3 = n + 5 - 3$, the balance is maintained. This also results in an equation that is easier to solve. For example,

$$2n + 3 = n + 5$$
$$2n + 3 - 3 = n + 5 - 3 \quad \text{(Subtract 3 from both sides.)}$$
$$2n = n + 2$$
$$n + n = n + 2 \quad \text{(Substitute } n + n \text{ for } 2n, \text{ since } 2n = n + n.\text{)}$$
$$n = 2$$

The equation representing the situation shown below on the balance scale on the left is $2b = 6$. Note that, in this model, the bags are closed and each contains the same number of cubes. To find out how many cubes are in each bag, each side could be separated into two equal amounts. This is equivalent to dividing both sides by 2, as shown by the balance scale on the right below

Dividing to Balance the Scale

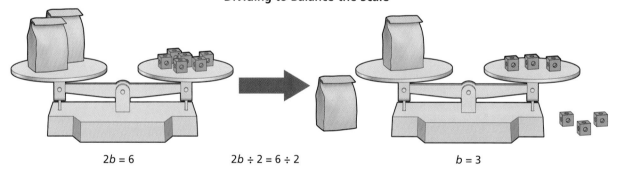

| $2b = 6$ | $2b \div 2 = 6 \div 2$ | $b = 3$ |

The equation representing the situation shown at the top of the next page is $2b = 6 + b$. To find out how many cubes are in each bag, a bag could be taken off each side. This is equivalent to subtracting b from both sides, as shown by the illustration.

Subtracting to Balance the Scale

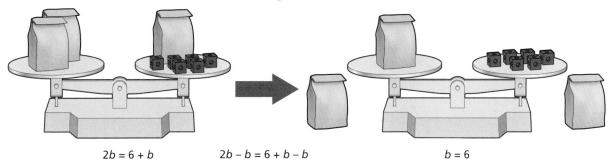

$2b = 6 + b$ $2b - b = 6 + b - b$ $b = 6$

Solving an Inequality

Students also should work informally with inequalities such as $3 + \square < 10$ or $3 + \square \neq 10$. At this level, students will likely guess and test to determine values that satisfy the inequality. They will observe that there are often, but not always, many possible whole number solutions.

INEQUALITY	NUMBER OF WHOLE NUMBER SOLUTIONS
$3 + \square < 10$	7 solutions: 0, 1, 2, 3, 4, 5, 6
$3 + \square \neq 10$	An infinite number of solutions: any number other than 7
$3 + \square < 3$	No whole number solutions

Using Graphs to Describe Relationships and Solve Equations

Graphs are effective models for describing and representing relationships between various variables or quantities visually. Some examples are shown below.

Graphs That Show Relationships

RELATING NUMBER OF TRICYCLES TO WHEELS

This graph shows a way for students to display the relationship between the number of tricycles and the total number of wheels; that is, as the number of tricycles increases by 1, the number of wheels increases by 3. By using the graph, the pattern inherent in counting tricycle wheels is highlighted.

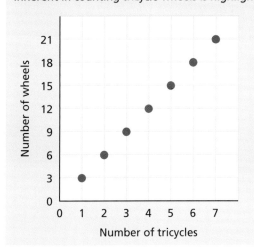

Number of tricycles

RELATING SIDE LENGTH TO PERIMETER

This graph shows that the perimeter of a square is related to its side length.

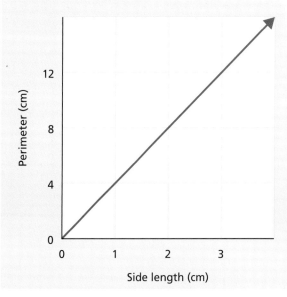

Side length (cm)

Graphs That Show Relationships (continued)

RELATING SIDE LENGTH TO AREA	RELATING MONTHS TO DAYLIGHT HOURS

This graph shows that the area of a square is related to its side length.

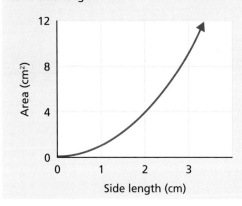

This graph shows that the length of daylight increases until June, and then decreases until the end of December.

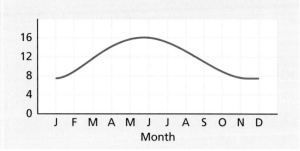

Teaching Idea | 1.14

Ask students to sketch a graph to represent the relationship between the amount of cereal in a bowl and time (BIA 1), if they start eating slowly and then speed up in order to get to school on time. Ask: *How does the graph show when the person is eating slowly? faster?* [by the steepness of the line] *Why is a graph a good way to show what happened?* [You can quickly tell what is happening by just looking.]

A graph of a relationship can be used to solve an equation as well. For example, to solve the equation $3x + 2 = 14$, a student can

- graph the relationship by plotting three points where the output value is 2 more than triple the input value, for inputs of 0, 1, and 2, and then drawing a line through the points
- look for an output, or y-value of 14 and then locate the related input, or x-value, which is 4

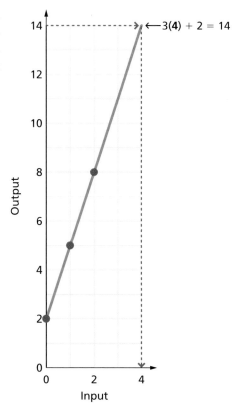

Chapter 2

Number and Operations

Numbers Beyond 1000

Students need strategies for representing and making sense of greater numbers. Although it is possible to count, say, 87 items individually, it is not practical. When items are grouped, counting is made easier and probably more accurate.

This becomes very clear when a teacher puts, for instance, 35 items in a disorganized fashion on an overhead projector, allows students to look for a few seconds, and then turns off the projector. Very few students can tell how many items there were. In contrast, if 35 items are shown in 3 groups of 10 and 5 more (as shown below), students find it easier to identify the amount. Grouping is a real help.

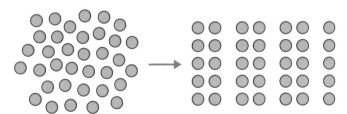

Once students start working with greater whole numbers, grouping becomes a necessary strategy.

BIG IDEAS FOR NUMBERS BEYOND 1000

1. The place value system we use is built on patterns to make our work with numbers efficient.

2. Students gain a sense of the size of numbers by comparing them to meaningful benchmark numbers.

Numeration Principles

Students who have explored the place value system for numbers up to 1000 need to extend this knowledge to work with greater numbers. Some of the principles that students need to learn are listed at the top of the next page.

A knowledge of big ideas can help teachers choose, shape, and create tasks and use questioning to help students make powerful connections.

Each teaching idea in this section of the chapter will indicate which Big Idea(s) for Numbers Beyond 1000 (BINBT) can be emphasized.

NUMERATION PRINCIPLES

1. Patterns are inherent in our numeration system because each place value is 10 times the value of the place to the right.
2. A number has many different "forms." For example, 12 145 is 1 ten thousand, 2 thousands, 1 hundred, 4 tens, and 5 ones, but it is also 121 hundreds and 45 ones.
3. A place value system requires a symbol for a place holder. For example, the 0s in 3004 are place holders. They push the digit 3 over to show that it represents thousands and not tens.
4. Numbers can be compared when written in standard, or symbolic, form.

The base ten place value system is defined by conventions that must be explicitly taught.

Teaching Idea | 2.1

Provide each student with seven cards with the words: forty, five, million, thousand, two, hundred, and six. Ask them to combine words to make numbers and to write each symbolically (e.g., forty-five million, six hundred thousand, two is 45 600 002).

Draw attention to BINBT 1 by asking: *How did you know your number would have seven or more digits?* [Two million is the least number and it has seven digits.] *How did you know the digit 4 would be the middle digit in a period?* [Forty means four tens, so it would have to be tens, ten thousands, or ten millions.]

Teaching Idea | 2.2

Presenting problems in contexts that involve numbers close to multiples of 100, 1000, and 10 000 will force transitions in counting patterns. For example,

The population of a town is 25 197. A new family moves into town. What might be the new population?

Focus on BINBT 1 by asking: *How did you know the hundreds, tens, and ones digits would change but not the thousands and ten thousands digits?* [I knew adding 3 would change the 7 to 0, then the 9 to 0, and finally the 1 to 2.]

PRINCIPLE 1 Patterns are inherent in our numeration system because each place value is 10 times the value of the place to the right.

Students at this level need to extend what they learned about relating ones to tens and tens to hundreds to recognize why the columns to the left of the hundreds are thousands, then ten thousands, then hundred thousands, and so on.

Although the names "thousands," "millions," and "billions" are conventions that must simply be transmitted to students, students can use their understanding of the structure of the place value system to see why the column to the left of thousands must be ten thousands and the column to the left of the ten thousands must be hundred thousands. The same would be true of ten millions and hundred millions or ten billions and hundred billions.

The Millions Period			The Thousands Period			The Ones Period		
1	2	3	1	3	5	1	5	7

123 135 157 is read as "123 million, 135 thousand, 157."

Students should discuss why grouping the digits of a numeral in threes, each called a period, helps us read and interpret large numbers more easily. For example, we can read the number 123 135 157 in three parts, as "123 million, 135 thousand, 157," since each period can be given a name.

Students can extend what they have learned using base ten blocks for numbers up to 1000 to gain insight into the place value system for larger numbers. For example,

From working with 3-digit numbers, they know

- the unit cube represents 1,
- 10 unit cubes make a rod representing 10,
- 10 rods side by side make a flat representing 100, and
- 10 flats in a stack make a large cube representing 1000.

From this, they can visualize

- 10 large cubes making a large rod representing 10 000,
- 10 large rods side by side making a large flat representing 100 000, and
- 10 large flats in a stack making a very large cube representing 1 000 000.

Students need considerable practice working with place value mats and base ten materials to internalize base ten place value counting patterns. Even older students have difficulty making transitions when they count with greater numbers. It is not unusual to hear a student count: "three thousand eight hundred ninety-eight (3898), three thousand eight hundred ninety-nine (3899), four thousand (4000)" rather than "three thousand nine hundred (3900)."

PRINCIPLE 2 A number has many different "forms."

The ability to rename numbers is fundamental to many of the algorithms involving addition, subtraction, multiplication, and division that students will learn. For example, regrouping 3003 as 300 tens + 3 ones makes it easy to see why $3003 \div 3 = 300$ tens + 3 ones $\div 3 = 100$ tens + 1 one, or 101.

When working with greater numbers, there are times when students will want to rename a number in different ways. For example, renaming 1 000 000 as 1000 thousands makes it easier to divide 1 000 000 by 1000, or renaming 1 000 000 as 10 hundred thousands makes it easier to divide it by 10 or 100 or 1000. Similarly, there are times when it is convenient to write 3 600 000 as 3.6 million, for example, to write a concise newspaper headline.

Students need many opportunities to represent a number in different ways, using materials like place value charts and base ten materials; using place value language, for example, 10 000 as 100 hundreds or 10 thousands; and using alternative forms based on computation, for example, 10 000 as 9999 + 1.

This ability will serve students well in many number situations. For example, students who understand why 10 000 is 100 hundreds will find it easier to understand why $1 \text{ m}^2 = 10\ 000 \text{ cm}^2$. Students who understand why 10 000 = 9999 + 1 will find it easier to use mental math to add 9999 to a number, by adding 10 000 and then subtracting 1.

Scientific Notation As students move toward high school, they learn to use the place value system in a new way to write numbers in scientific notation. The purpose of this notation is to write the number in a concise way in order to convey a quick sense of the size of a number. For example, if 4236 is written as 4.236×10^3, it is quickly apparent that the number is somewhere in the 1000s range (1000 to 9999) because $1 \times 10^3 = 1000$ and $1 \times 10^4 = 10\ 000$. The digit 4 representing 4000 in 4236 is the most significant digit in the number 4236 (of all four digits, it alone conveys a good sense of the size of the number). That is why it is the only digit to the left of the decimal point when 4236 is written in scientific form.

Converting numbers in standard form to scientific notation and vice versa emphasizes and requires an understanding of the pattern of the place value system. For example, 123 000 becomes 1.23×10^5 because the most significant digit (the digit in the place representing 10^5) moves over 5 places to the ones place. Similarly, 3.46×10^7 becomes 34 600 000 because 3.46×10^7 is about 3 ten millions, so the digit 3 must move over 7 places to the ten millions place.

PRINCIPLE 3 A place value system requires a symbol for a place holder.

When you write a number in its symbolic standard form using digits, for example, 3004, the digit 0 is a place holder. Many people know the term place holder, but have difficulty explaining what it means. The idea is that if you did not have the digit 0, the number would be recorded as 34, and you would mistakenly think that the 3 represented 30 instead of 3000.

Teaching Idea | **2.3**

Pose tasks like these to focus students on different ways to represent 1 million:

- How many loonies is 1 million worth? How many toonies? How many $100 bills?
- How long would a line of 1 million pennies be?
- How long would the side of the square made of 1 million pennies be?
- How long would it take to roll 1 million pennies?
- How high would a stack of 1 million pennies be?

Focus on BINBT 2 by asking: *What benchmark numbers did you use to solve the problem?* [e.g., 100; I figured out how long it would take to roll 100 pennies and then multiplied by 10 000.]

Teaching Idea | **2.4**

Ask students to use the digits 1, 3, 5, 6, and 7 to complete ■.■■■ × 10■ to create 10 numbers, including the least and greatest numbers possible, and to write each number in standard form.

Draw attention to BINBT 1 by asking: *How did you arrange the digits to create the greatest number?* $[6.531 \times 10^7]$ *the least number?* $[3.567 \times 10^1]$

When you represent a number using a place value mat or with base ten blocks, the digit 0 is not necessary, as shown below. However, in symbolic recordings, you need the digit 0 as a place holder.

Thousands	Hundreds	Tens	Ones
⬤ ⬤ ⬤			⬤ ⬤ ⬤ ⬤

Representing 3004 on a place value mat

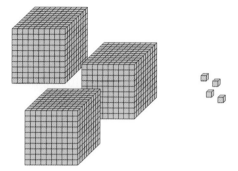

Representing 3004 with base ten blocks

When modelling 3004 concretely, you do not need a place holder for zero.

Teaching Idea | **2.5**

Present students with riddles like this:

I'm thinking of a number.

- It is between 20 000 and 60 000.
- Each digit is even.
- The sum of the digits is 10.

What is the number? Find more than one answer.

Students can create riddles for other students.

Draw attention to BINBT 1 by asking: *Why might a clue for the number 1 000 000 be "It has one more digit than 100 000 has"?* [100 000 has 6 digits, so it has to be a number with 7 digits, which is between 1 000 000 and 9 999 999.]

PRINCIPLE 4 Numbers can be compared when written in standard, or symbolic, form.

Once whole numbers are represented in their standard, or symbolic, form students can use the number of digits to get a sense of the size of the numbers in order to compare them. For example, any 3-digit whole number is 100 or greater, but less than 1000. Similarly, any 2-digit whole number is less than 100. Therefore, any 3-digit whole number is automatically greater than any 2-digit whole number.

When the numbers have the same number of digits, students can compare numbers by thinking about the role of the various digits in a numeral in order to realize that there are digits that are more and less important in the numeral. So, when two whole numbers have the same number of digits, the left-most digits matter most when ascertaining the size of the numbers, because that place has the greatest value. For example, in comparing 3021 and 5974, the 3 and the 5 are quite important because, of all the digits, they alone provides a sense of the relative size of the two numbers. The digits 1 and 4 have some importance, perhaps, in clarifying that the number is odd or even, but none in terms of the size of the numbers.

Comparing Numbers	
A DIFFERENT NUMBER OF DIGITS	**THE SAME NUMBER OF DIGITS**
42 395 > 4172 because 42 395 > 10 000 but 4172 < 10 000	42 395 > 31 399 because 40 000 > 30 000

Estimating Numbers

Often you do not require an exact amount to represent a number, only an estimate. Students should be comfortable with a variety of ways of estimating numbers, depending on the situation and the numbers involved.

Estimating is used in situations such as these:

- In computation; for example, to estimate $3482 + 2712$ you could round 3482 to 3500 and 2712 to 3000 to estimate $3500 + 3000 = 6500$.
- To get a sense of the size of numbers in order to compare them; for example, 3826 is a bit more than 3800 and 3159 is a bit less than 3200, so $3826 > 3159$.
- For reporting numbers; for example, about 20 000 people came to an event.

Rounding to Estimate Numbers

One approach to estimating is to follow the conventions for rounding a number to the nearest multiple of 10, 100, 1000, 10 000, 100 000, ... To round a number, students can think of a number line as a highway with "gas stations" at the multiples to which the number could be rounded. The gas station metaphor provides a context where students can think, "I am almost out of gas; which gas station does it make more sense for me to drive to?" The number to round to is the location of that "gas station." For example, when rounding to the nearest thousand, the gas stations are at multiples of 1000. If you are at 13 427, it makes sense to drive to 13 000 because it is closer than 14 000. However, if you are at 13 500, it makes more sense to go to 14 000 because, even though 13 000 and 14 000 are an equal distance away, going to 14 000 is the better choice—it will take you farther along your route.

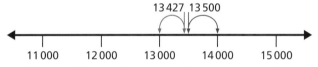

Following the rounding conventions, 13 427 rounds to 13 000, while 13 500 rounds to 14 000.

Benchmark Numbers

It is important to stress that the rounding conventions described above are only one way to round numbers. There could be situations where it might make more sense to round 13 427 to 13 425 because of the numbers involved. For example, if you were estimating the sum of $13\,427 + 72$, you might add $13\,425 + 75 = 13\,500$.

Numbers that you round to are often called benchmark, anchor, or "comfortable" numbers. Many people are comfortable with the 25s, likely because of the monetary system, so rounding to multiples of 25 often makes sense. For example, a principal might estimate a class that has between 23 and 28 students as having about 25 students instead of estimating it as about 20 or 30.

To become comfortable with greater benchmark numbers, students might enjoy exploring numbers like 1 million or 1 billion in contextually interesting ways. You might want to use some of the David Schwartz books about a million to introduce some of these activities.

Teaching Idea | **2.6**

Students can build two base ten block number towers with the same height as this one and then write the numbers symbolically.

Focus on BINBT 1 and 2 by asking: *How do you know each number will be of the form* 13■ ■? [Each tower has more than 1 thousand cube and 3 hundred flats (1300) and less than 1 thousand cube and 4 hundred flats (1400).]

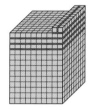

A base ten block tower for 1320

The convention for rounding numbers to the nearest 10, 100, 1000, and so on, is only one way to round numbers.

Number Theory

Number theory is the study of multiplicative and divisive properties of numbers.

Number theory is the study of integers, often focusing on whole numbers, that deals with multiplicative and divisive properties of numbers, as opposed to number representation in the place value system.

Multiplicative thinking focuses on how one number is made up of groups of another "unit," as opposed to additive thinking. For example, thinking of 20 as 4 fives is thinking of 20 in groups of 5. In contrast, thinking of 20 as $11 + 9$ is additive thinking.

As students work with multiplication and division, they deal with the concepts of divisibility, multiples, and factors.

Each teaching idea in this section of the chapter will indicate which Big Idea(s) for Number Theory (BINT) can be emphasized.

BIG IDEAS FOR NUMBER THEORY

1. Thinking of numbers as factors or multiples of other numbers provides alternative representations of those numbers.

2. Classifying numbers as factors and/or multiples of other numbers, or as primes or composites, provides additional information about those numbers.

3. Just as multiplication and division are intrinsically related, so are factors and multiples.

Multiples and Factors

A multiple of a number is the product of that number with an integer multiplier. For example, 8 is a multiple of 2 and a multiple of 4 since $8 = 2 \times 4$ and both 2 and 4 are integers. Note that even though 9 can also be expressed as the product of 4 and another number ($9 = 2.25 \times 4$), it is not a multiple of 4 since its multiplier is not an integer.

Whenever there is a multiple, there are factors. The number 18 is a multiple of 6 because $3 \times 6 = 18$, which means 18 is also a multiple of 3. Therefore, 3 and 6 are factors of 18.

Students will discover that some numbers have many factors, some have a few, and some have only one or two. For example,

FACTORS OF 24		FACTORS OF 6		FACTORS OF 97	
1	24	1	6	1	97
2	12	2	3		
3	8				
4	6				
8 factors: 1, 2, 3, 4, 6, 8, 12, 24		4 factors: 1, 2, 3, 6		2 factors: 1, 97	

As shown in the chart above, factors come in pairs, although some numbers have an odd number of different factors (square numbers such as 16), and the number 1 has only one factor (it is the only such counting number) as shown at the top of the next page.

Teaching Idea | **2.7**

Assign this activity to focus on BINT 1, 2, and 3.

A 2-digit number is a multiple of 4. Tell everything you know about the number.

Follow up by asking: *How do you know it's even?* [A multiple of 4 is a multiple of 2.] *How do you know it has a factor of 4?* [It's a multiple of 4.] *How do you know it might or might not be a multiple of 3?* [Some 2-digit multiples of 4 are also multiples of 3, like 12, 24, 36, …, but some are not, like, 4, 8, 16, ….]

FACTORS OF 16		FACTORS OF 1	
1	16	1	1
2	8		
4	4		
5 factors: 1, 2, 4, 8, 16		1 factor: 1	

Determining Factors

Organized lists, like those above, are one way of determining factors in a systematic fashion, beginning with 1 and the number itself, and then 2 or the next possible factor and its factor partner, and so on. Another way to organize and display the factors of a number is using a factor rainbow. For example,

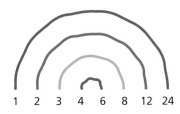

A factor rainbow for 24

If students construct a factor rainbow for a square number like 100, they will discover a repeated factor (10), which only needs to be listed once.

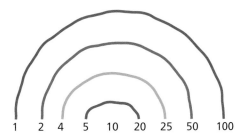

A factor rainbow for the square number 100

Concrete models and other pictorial models can also be used. For example, to determine the factors of 12, take 12 square tiles and try to arrange them into a rectangle. Record the length and width of each rectangle you can make; these are the factor pairs. This can also be done pictorially by drawing rectangles on grid paper (as shown below). You can use the same strategy as an organized list by approaching this systematically, beginning with a width of 1 unit, and then 2 units or the next possible width that is a factor, and so on.

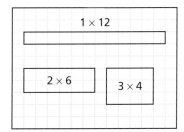

Factors of 12: 1, 2, 3, 4, 6, 12
Creating rectangles to find factors

Teaching Idea | 2.8

Ask students to look at the numbers in a factor rainbow for several even and odd numbers.

To focus on BINT 2 (specifically how classifying a number as a multiple of 2 or as an even number also tells you something about its other factors), ask: *When is the second number from the right exactly half of the number farthest right?* [when the number is even] *When is it not?* [when the number is odd] *What does this tell you about the second greatest factor of an even number?* [It's always half of the number.]

Factors and Divisibility

If a number is divisible by another number, it means the first number is a multiple of the second number (the second number being a factor). There are a number of tests for determining if a number is a multiple of certain factors. Some of the tests are described below.

DIVISIBILITY TESTS

A number is DIVISIBLE BY 2 if the ones digit is 0, 2, 4, 6, or 8.
This works because of the pattern of the number system and because 2 is divisible by itself. Every second number follows the pattern 2, 4, 6, 8, x0, x2, x4, x6, x8, x0, with only the digits 0, 2, 4, 6, and 8 appearing in the ones place.

A number is DIVISIBLE BY 3 if the sum of the digits is a multiple of 3.
Consider the number 414 (4 hundreds, 1 ten, and 4 ones):

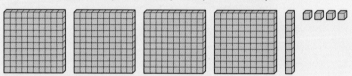

- If 1 hundred block is divided into groups of 3, there will be 1 one block left over. So, if the 4 hundred blocks are grouped into 3s, there will be 4 one blocks left over.
- If the 1 ten block is grouped into 3s, there will be 1 one block left over.
- There were 4 one blocks to start with.
- After grouping the 4 hundreds and 1 ten into as many 3s as possible, there are 9 one blocks left over (since 4 + 1 + 4 = 9), and 9 can be grouped into 3s because it is a multiple of 3.

This explains why 414 is a multiple of 3 and holds true for any number for which the sum of the digits is a multiple of 3.

A number is DIVISIBLE BY 4 if 2 × the tens digit + the ones digit is a multiple of 4.
Every multiple of 100, 1000, 10 000, and so on can divided by 4 with no remainder. So only the tens and ones digits have to be considered.
If each 10 is broken up into 8 + 2, the 8 part is divisible by 4, so only the 2 part needs to be considered.
For example,
352 = 3 × 100 + 5 × 10 + 2 = 3 × 100 + 5 × 8 + 5 × 2 + 2
3 × 100 is a multiple of 4, and so is 5 × 8. Since 5 × 2 + 2 is also a multiple of 4, 352 is a multiple of 4.

A number is DIVISIBLE BY 5 if the ones digit is 5 or 0.
This works because 5 × 1 = 5. Further multiples are determined by adding 5. When you add 5 to any number that has 5 as the ones digit, the ones digit is 0. When you add 5 to any number that has 0 as the ones digit, the ones digit is 5.

A number is DIVISIBLE BY 6 if it is even (or divisible by 2) and divisible by 3.
The multiples of 3 are 3, 6, 9, 12, 15, 18, ...
Every second multiple of 3 is even (6, 12, 18, ...); these are the multiples of 6.

A number is DIVISIBLE BY 11 if the difference between sums of alternate digits is a multiple of 11.
The powers of 10 alternate between being 1 greater and 1 less than a multiple of 11.
1 = 0 + 1   10 = 11 − 1   100 = 99 + 1
1000 = 1001 − 1   10 000 = 9999 + 1   100 000 = 100 001 − 1
Because of this, if you add the digits in the 1s, 100s, 10 000s, ... places, and then subtract the sum of the digits in the 10s, 1000s, 100 000s, ... places, the difference is a multiple of 11 (including 0) if the number is a multiple of 11.
For example, 5412 is divisible by 11 since 2 + 4 − (1 + 5) = 0 and 979 is divisible by 11 since 9 + 9 − (7 + 0) = 11.

Teaching Idea | **2.9**

Ask students to create a divisibility test for 12 or 15.

Bring attention to BINT 2 by asking: *Why does knowing a number is divisible by 12 (or 15) also tell you it is divisible by 3?* [If a number is divisible by 12, you can divide it into groups of 12. You can then divide each group of 12 into groups of 3.] *How might that be useful in creating a divisibility test for 12 or 15?* [Since numbers divisible by 12 and 15 are divisible by 3, part of the test will be that the sum of the digits is a multiple of 3.]

Teaching Idea | **2.10**

Provide practice with divisibility by having students play this game.

Each player chooses three cards from a deck and arranges them into a number that is divisible by as many of the factors 2, 3, 4, 5, 6, 8, 9, and 10 as possible. Points are scored according to the number of factors.

Emphasize BINT 2 by asking: *How do you know you will get at least three points if your ones digit is 0?* [It will be divisible by 10, 5, and 2.]

Prime and Composite Numbers

Numbers with exactly two factors have a special name—prime numbers, or primes. There are very small prime numbers, such as 2, 3, and 5, but there are also large ones, such as 6299. In fact, mathematicians continue to identify greater and greater prime numbers. Every whole number other than 1 that is not a prime is called a composite number, and has three or more factors. Note that 2 is the only even prime number and the number 1 is neither prime nor composite (mathematicians call it a unit).

Determining Whether a Number Is Prime

There are many ways to decide whether a number is prime. One way, which is very tedious, is to try to divide the number by every possible smaller number to see how many factors it has. An interesting way to find the prime numbers between 1 and 100 is to use a technique called the Sieve of Eratosthenes, as described below.

Finding Prime Numbers from 1 to 100

THE SIEVE OF ERATOSTHENES

Use a 100 chart and coloured counters:
Step 1 Place a blue counter on 1.
Step 2 Place red counters on every multiple of 2 but not 2 itself.
Step 3 Place yellow counters on every uncovered multiple of 3 but not 3 itself.
Step 4 Place blue counters on uncovered multiples of 5 but not 5 itself.
Step 5 Place green counters on uncovered multiples of 7 but not 7 itself.
The uncovered numbers are primes: 2, 3, 5, 7, 11, 13, 17, 19, 23, 29, 31, 37, 41, 43, 47, 53, 59, 61, 67, 71, 73, 79, 83, 89, and 97.

The Sieve of Eratosthenes "catches" the prime numbers.

Some students will observe that the only possible final digits for primes greater than 2 and 5 are 1, 3, 7, and 9. This occurs because no multiple of 2 (other than 2) can be prime (since it has a minimum of three factors: 1, 2, and itself), and no multiple of 5 other than 5 itself can be prime (since it has a minimum of three factors: 1, 5, and the number itself).

At some point, students will learn that the primes are the "building blocks" of our whole number system in the sense that each whole number can be broken down into prime factors in one unique way. This is called prime factorization; for example, $36 = 2 \times 2 \times 3 \times 3$. Students often use factor trees to determine these prime factors. Notice below that there is often more than one factor tree for any given number, but they all end up with the same list of prime factors.

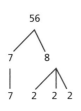
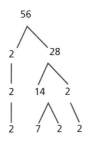

Using a factor tree to find prime factors for 56 ($2 \times 2 \times 2 \times 7$)

Students can find the primes on a 100 chart with 6 columns instead of 10. They will discover that classifying a number greater than 3 as prime tells you something else about the number (BINT 2). Elicit this by asking: *In which column did you find multiples of 6?* [sixth or last column] *the primes?* [first and fifth columns] *What do you notice?* [Primes are 1 more or 1 less than a multiple of 6.]

1	2	3	4	5	6
7	8	9	10	11	12
13	14	15	16	17	18

There are many famous conjectures about primes that demonstrate BINT 2. Ask students whether they think each conjecture is true.

- Every even number but 2 is the sum of two primes; e.g., $6 = 3 + 3$ and $8 = 5 + 3$. (Goldbach's conjecture)
- Every odd number is the sum of a prime and a power of 2; e.g., $15 = 7 + 2^3$ and $83 = 2 + 3^4$.

[Both are considered true.]

Then ask: *What other conjectures about primes could you create and test?* [e.g., There is always one or more prime number between consecutive square numbers; e.g., 7 is between 2^2 and 3^2 and 101 is between 10^2 and 11^2.]

Factors and Multiples

Common Factors and GCF

Sometimes you need to determine factors that two numbers have in common. Common factors can be used to solve many mathematical problems such as the one shown in **Teaching Idea 2.13**. They are also useful for expressing fractions in lowest or simplest terms.

Teaching Idea | **2.13**

Pose the following problem:

A room that is 20 m by 12 m is to be tiled with square tiles. If the tiles must fit perfectly without cutting and be a whole number of metres on each side, what is the largest tile size that could be used? [4 m by 4 m]

Focus on BINT 1 by asking: *How did you use common factors to solve the problem?* [4 is the GCF of 20 and 12.]

FINDING COMMON FACTORS AND GCF	USING THE GCF
Factors of 18: <u>1</u>, <u>2</u>, <u>3</u>, <u>6</u>, 9, 18, Factors of 30: <u>1</u>, <u>2</u>, <u>3</u>, 5, <u>6</u>, 10, 15, 30 Common factors: 1, 2, 3, and 6 Greatest common factor (GCF): 6	To write the fraction $\frac{18}{30}$ in lowest terms, divide the numerator and denominator by the greatest common factor (GCF): GCF (18, 30) = 6, so $\frac{18}{30} = \frac{18 \div 6}{30 \div 6} = \frac{3}{5}$

Common Multiples and LCM

Sometimes you need to determine the multiples that two numbers have in common. Common multiples come in handy for solving problems like the one shown in **Teaching Idea 2.14** and for later work with fractions to determine the least common denominator (which is the least common multiple of the numbers in the denominators).

Teaching Idea | **2.14**

Pose problems like this that involve common multiples or factors to focus on BINT 1:

Kyle bought some $12 shirts. Art bought some $15 shirts. They both spent the same amount, which was less than $200. How much could they have spent? [$60, $120, $180]

Extend the problem by asking: *How would your answer change if Art's shirts cost $16?* [$48, $96, $144, $192] *Why?* [Common multiples of 12 and 16 are multiples of 48.]

FINDING COMMON MULTIPLES AND LCM	USING THE LCM
Multiples of 12: 12, 24, 36, 48, <u>60</u>, 72, 84, 96, 108, <u>120</u>, ... Multiples of 15: 15, 30, 45, <u>60</u>, 75, 90, 105, <u>120</u>, ... There is an infinite number of common multiples of 12 and 15: 60, 120, 180, ... Lowest common multiple (LCM): 60	To write the fractions $\frac{7}{12}$ and $\frac{6}{15}$ using the lowest common denominator (LCD), use the lowest common multiple (LCM) of the denominators 12 and 15. LCM (12, 15) = 60, so $\frac{7}{12} + \frac{6}{15} = \frac{35}{60} + \frac{24}{60}$

Exponents

Teaching Idea | **2.15**

To help students focus on alternative representations of numbers and the relationship between factors and multiples (BINT 1 and 3), ask: *A number is a power of 4. Can it also be a power of 2? How do you know?* [yes, 4 = 2 × 2 so, e.g., 4 × 4 × 4 = 2 × 2 × 2 × 2 × 2 × 2] *If a number is a power of 2, can it also be a power of 4? Explain.* [Only if the exponent is even; e.g., 2^8 = 2 × 2 × 2 × 2 × 2 × 2 × 2 × 2 = 4 × 4 × 4 × 4 = 4^4, but 2^7 = 2 × 2 × 2 × 2 × 2 × 2 × 2 = 4 × 4 × 4 × 2 = 4^3 × 2]

Writing numbers using exponents is representing a number in the unit of the "base." For example, $8 = 2 \times 2 \times 2$, so we write $8 = 2^3$. In effect, we are using number theory ideas since we are thinking of 8 in units of 2. We say that 8 is "the third power of 2" and 2^3 is "2 to the third."

3 is the exponent to which the base of 2 is raised.

Using a power is shorthand for multiplying, just as using multiplication is shorthand for adding. 2^3 is quicker to write than $2 \times 2 \times 2$, just as 3×2 is quicker to write than $2 + 2 + 2$.

Whole Number Operations

In Grades 4 to 8, while students solidify their understanding of addition and subtraction (the focus in K to 3) they focus on developing their understanding of multiplication and division further. This includes knowing when to multiply and when to divide, and an ability to use a variety of multiplication and division strategies.

BIG IDEAS FOR WHOLE NUMBER OPERATIONS

1. There are many situations to which an operation is applied and there are many procedures, or algorithms, for each operation.

2. A personal "invented" algorithm is often more meaningful and sometimes equally efficient as a conventional algorithm.

3. Operation procedures should be taught meaningfully, taking into account the various meanings of the operations and the principles that apply to their use. This is facilitated by concrete manipulations and pictorial models.

4. There are a variety of appropriate ways to estimate sums, differences, products, and quotients depending on the numbers involved and the context.

Each teaching idea in this section of the chapter will indicate which Big Idea(s) for Whole Number Operations (BIWNO) can be emphasized.

Multiplication and Division Meanings and Principles

Meanings Across the Grades

MULTIPLICATION IN KINDERGARTEN TO GRADE 3	DIVISION IN KINDERGARTEN TO GRADE 3
Students are introduced to these meanings: • repeated addition For example, $3 + 3 + 3 + 3 = 4 \times 3 = 12$ • the total count of a set of equal groups For example, 4 groups of boys with 3 in each is 12 boys: $4 \times 3 = 12$ • the total count in an array For example, An array of 4 rows with 3 stamps in each is 12 stamps: $4 \times 3 = 12$	Students are introduced to these meanings: • repeated subtraction For example, $12 - 4 - 4 - 4 = 0 \rightarrow 12 \div 4 = 3$ • the number of groups of a given size in a total For example, 12 boys grouped in 4s is 3 groups: $12 \div 4 = 3$ • the size of a share if a total is shared equally For example, 12 boys put into 4 groups is 3 boys in each group: $12 \div 4 = 3$ • the number of rows or columns in an array For example, An array of 12 stamps in 4 rows has 3 columns: $12 \div 4 = 3$ (continued)

Pose combination problems like this:

There are two types of bread (white and brown) and three types of cheese (cheddar, Swiss, and mozzarella). How many ways can you combine them to make different sandwiches?

Help students see this as a multiplication situation (BIWNO 1) by asking: *How could you show the different bread–cheese combinations as equal groups?* [e.g., 1 group with 3 sandwiches: W-C, W-S, W-M, and another group with 3 sandwiches: B-C, B-S, B-M]

The recognition of these additional meanings allows students to solve a broader range of problems. Recognizing the relationship between rate and multiplication and between division and fractions is particularly important in Grades 6 to 8.

Meanings Across the Grades (continued)

MULTIPLICATION IN GRADES 4 TO 8	DIVISION IN GRADES 4 TO 8

In Grades 4 to 8, students extend their understanding of multiplication to include these meanings:

Combinations

3×4 is the number of paper–envelope combinations possible, if there are 3 kinds of envelopes and 4 colours of paper.

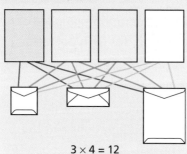

$3 \times 4 = 12$

3 envelopes and 4 colours of paper makes 12 different paper-envelope combinations.

Rates

3×4 is the number of pencils Poli has, if Mia has 4 pencils and Poli has 3 times as many.

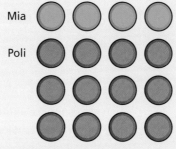

Mia

Poli

$3 \times 4 = 12$
At a rate of 3 times as many, Poli has 12, if Mia has 4.

In Grades 4 to 8, students extend their understanding of division to include these meanings:

Width or length of a rectangle

$12 \div 4$ is the width of a rectangle that has an area of 12 square units and a length of 4 units.

Area is 12 square units

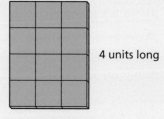

4 units long

$12 \div 4 = 3$ units wide

Fractions

$12 \div 3$ is the number of items you have when you divide 12 items into thirds.

4 is $\frac{1}{3}$ of 12.

Multiplication and Division Principles

Students need to become familiar with and use a number of principles about multiplication and division. In the explanations of the principles on the following pages, different meanings of multiplication and different models are used each time. Often, one meaning of an operation is much more helpful than another in explaining a particular idea.

MULTIPLICATION AND DIVISION PRINCIPLES

1. Multiplication and division "undo" each other. They are related inverse operations, e.g., if $12 \div 3 = 4$, then $3 \times 4 = 12$.

2. You can multiply numbers in any order (the commutative property). However, with division, the order in which you divide the numbers matters.

3. To multiply two numbers, you can divide one factor and multiply the other by the same amount without changing the product (the associative property), e.g., $8 \times 3 = (8 \div 2) \times (3 \times 2) = 4 \times 6$.

4. To divide two numbers, you can multiply or divide both numbers by the same amount without changing the quotient, e.g., $15 \div 3 = (15 \times 2) \div (3 \times 2) = 30 \div 6$.

5. You can multiply in parts (the distributive property), e.g., $5 \times 4 = 3 \times 4 + 2 \times 4$.

6. You can multiply in parts by breaking up the multiplier, e.g., $6 \times 5 = 2 \times 3 \times 5$.

7. You can divide in parts by splitting the dividend into parts, but not the divisor (the distributive property), e.g., $48 \div 8 = 32 \div 8 + 16 \div 8$.

8. You can divide by breaking up the divisor, e.g., $36 \div 6 = 36 \div 3 \div 2$.

9. When you multiply by 0, the product is 0.

10. When you divide 0 by any number but 0, the quotient is 0.

11. You cannot divide by 0.

12. When you multiply or divide a number by 1, the answer is the number you started with.

Some properties have been named by mathematicians, in particular, the commutative, associative, and distributive properties (as identified in the list).

PRINCIPLE 1 Multiplication and division "undo" each other. They are related inverse operations.

Multiplication and division are inverse operations; one "undoes" the other. In fact, division is defined as the "opposite" of multiplication. If you start with 12 items and share them among 3 people, each person gets 4 items. How do you get back to where you started (to 12)? You think of the 3 people, each with 4 items, as a multiplication situation, or 3 groups of 4, which is 12.

PRINCIPLE 2 You can multiply numbers in any order (the commutative property). However, with division, the order in which you divide the numbers matters.

If you model 3 groups of 4 as shown below, it is not clear why it is the same as 4 groups of 3.

Teaching Idea | **2.17**

To focus students on the relationship between multiplication and division (BIWNO 1), ask: *I am sharing 24 cookies with three other people. How could I multiply to figure out the number of cookies I would get?* [Since $4 \times 6 = 24$, then $24 \div 4 = 6$.]

3 groups of 4 is $3 \times 4 = 12$.

But, if you arrange 12 items in an array, it is obvious why $3 \times 4 = 4 \times 3$.

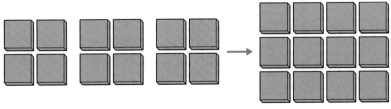

3 groups of 4 is $3 \times 4 = 12$
3 rows of 4 = 12 $3 \times 4 = 12$
4 columns of 3 = 12 $4 \times 3 = 12$

Modelling to explain the commutative property of multiplication

The order does matter when you divide, however. For example, if you were to divide 12 items into groups of 3, you would have 4 groups of 3. If you were to divide 3 items into groups of 12, you would not have even 1 full group. So, $12 \div 3 \neq 3 \div 12$.

PRINCIPLE 3 To multiply two numbers, you can divide one factor and multiply the other by the same amount without changing the product (the associative property). The associative property of multiplication suggests that you can multiply $a \times b \times c$ by multiplying $a \times b$ and then multiply the product by c, or you can multiply $b \times c$ and then multiply by a. For example, $4 \times 2 \times 3 = 8 \times 3$ or 4×6. The property exists since multiplication is only defined for two numbers, and rules had to be created to allow you to deal with more numbers than that.

Consider the 8 groups of 3 (8×3) shown below. If you pair up groups of 3, you will have 6 in each group (twice as many in each group) but only 4 groups (half as many groups), while the total number of circles stays the same ($8 \times 3 = 4 \times 6$).

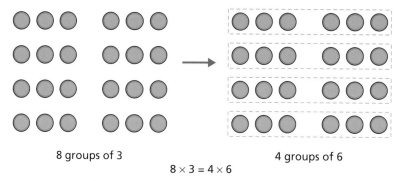

8 groups of 3   4 groups of 6
$8 \times 3 = 4 \times 6$

8 groups of 3 can be viewed as 4 groups of 6 (a consequence of the associative property of multiplication).

Teaching Idea | **2.18**

To focus students on the many ways we can divide (BIWNO 1), have them solve this problem three different ways:

There are 32 markers and 8 groups. How many markers will each group get? [e.g., $32 \div 8 = ? \rightarrow 8 \times ? = 32 \rightarrow 8 \times 4 = 32$, so $32 \div 8 = 4$; $32 \div 8 = 16 \div 4 = 4$; $32 \div 8 = 32 \div 2 \div 2 \div 2 = 4$]

PRINCIPLE 4 To divide two numbers, you can multiply or divide both numbers by the same amount without changing the quotient.

For instance, $15 \div 3$ asks how much each person gets if 3 people share 15 items equally. It makes sense that if there are twice (or half) as many items to be shared by twice (or half) as many people, the share size stays the same. It does not matter what the dividend and divisor are multiplied (or divided) by. As long as they are multiplied (or divided) by the same amount, the quotient does not change.

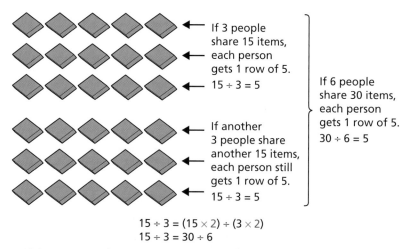

If 3 people share 15 items, each person gets 1 row of 5.
$15 \div 3 = 5$

If another 3 people share another 15 items, each person still gets 1 row of 5.
$15 \div 3 = 5$

If 6 people share 30 items, each person gets 1 row of 5.
$30 \div 6 = 5$

$15 \div 3 = (15 \times 2) \div (3 \times 2)$
$15 \div 3 = 30 \div 6$

If there are twice as many items to be shared by twice as many people, the share size stays the same.

PRINCIPLE 5 You can multiply in parts (the distributive property).

For instance, you can separate 5 rows of 4 squares into 3 rows of 4 and 2 rows of 4 ($5 \times 4 = 3 \times 4 + 2 \times 4$) without changing the total number of squares.

Teaching Idea | **2.19**

Create a 5×7 array and then divide it into groups of rows or columns (see Principle 5).

To focus on alternative strategies for multiplying (BIWNO 1), each time, ask students what two products could be added to solve 5×7 [e.g., $5 \times 3 + 5 \times 4$ or $2 \times 7 + 3 \times 7$], then ask: *How would separating the array this way help you multiply 5×7?* [If I didn't know 5×7, I could use 5×3 and 5×4.] *What else could you do if you only knew the doubles of 5?* [$5 \times 7 = 2 \times 5 + 2 \times 5 + 1 \times 5$]

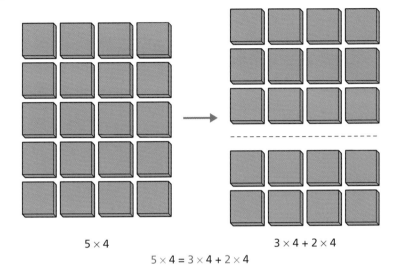

5×4 $3 \times 4 + 2 \times 4$

$5 \times 4 = 3 \times 4 + 2 \times 4$

Distributing 5 rows of 4 over 3 rows of 4 and 2 rows of 4

PRINCIPLE 6 You can multiply in parts by breaking up the multiplier.

An array is a good way to show this principle. For example, you can easily separate 6 rows of 5 squares into 2 groups, each with 3 rows of 5, without changing the total number of squares ($6 \times 5 = 2 \times 3 \times 5$).

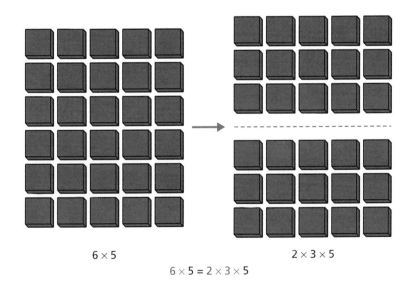

6×5 $2 \times 3 \times 5$

$6 \times 5 = 2 \times 3 \times 5$

An array of 6 rows of 5 can be viewed as 2 arrays, each with 3 rows of 5.

PRINCIPLE 7 You can divide in parts by splitting the dividend into parts, but not the divisor (the distributive property).

For example, $48 \div 8$ asks how many items each person gets when 48 items are shared equally. It is possible to "distribute" the first 24 among the 8 people, and then "distribute" the other 24 among the same 8 people ($48 \div 8 = 24 \div 8 + 24 \div 8$), hence the name: the distributive property (as shown on the next page).

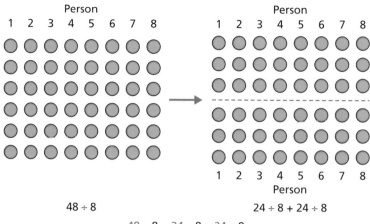

Person
1 2 3 4 5 6 7 8

Person
1 2 3 4 5 6 7 8

1 2 3 4 5 6 7 8
Person

48 ÷ 8

24 ÷ 8 + 24 ÷ 8

$$48 \div 8 = 24 \div 8 + 24 \div 8$$

Sharing 48 among 8 people is the same as sharing the first 24 among 8 people, and then sharing the remaining 24 among the same 8 people.

PRINCIPLE 8 You can divide by breaking up the divisor.

The expression $36 \div 6$ tells how many items there are in each group if 36 items are shared equally among 6 groups. You could share the 36 items among 6 groups ($36 \div 6$), or you could split the 36 items into 2 groups of 18 and then share each group of 18 into 3 smaller groups of 6 ($36 \div 6 = 36 \div 2 \div 3$).

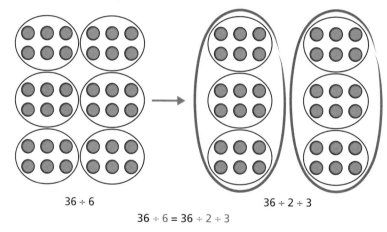

36 ÷ 6

36 ÷ 2 ÷ 3

$$36 \div 6 = 36 \div 2 \div 3$$

Share 36 into 6 groups by sharing first into the 2 groups circled in red, and then sharing within each red group into 3 groups.

PRINCIPLE 9 When you multiply by 0, the product is 0.

To show, say, 5 sets of 0, you might use 5 empty plates. Since there is nothing on any of the plates, the total number of items is 0 ($5 \times 0 = 0$).

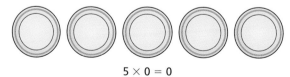

$5 \times 0 = 0$

It does not matter how many empty plates there are, any number of plates with 0 items on them results in 0 items altogether.

PRINCIPLE 10 When you divide 0 by a number other than 0, the quotient is 0.
For example, $0 \div 5$ asks the share size if 5 people share nothing. The amount is clearly nothing ($0 \div 5 = 0$).

Teaching Idea | **2.20**

Have students figure out the share size if 36 items are shared equally by 6 people. Help them see that they could

• divide the 36 items into 6 groups, or

• divide the 36 items into 2 groups and then divide each group into 3 groups, or

• divide the 36 items into 3 groups and then divide each group into 2 groups

To help students see the value of alternative strategies for dividing (BIWNO 1), ask: *Why might different students use these different strategies?* [e.g., dividing 36 by 2 and then 18 by 3 might be easier for them than dividing 36 by 6]

PRINCIPLE 11 You cannot divide by 0.

How can a number of items be shared among 0 groups? Here a repeated subtraction meaning for division might help explain this principle. To divide, for example, 15 ÷ 5, you can subtract 5 three times from 15, until you get to 0, so 15 ÷ 5 = 3. So, to divide 15 ÷ 0, you must determine how many times you can subtract 0 from 15 before getting to 0. There is no answer because you will simply never get to 0. Therefore, 15 ÷ 0 is undefined. As for 0 ÷ 0, you can subtract 0 any number of times from 0 to get to 0, so the answer could be 1, 2, 3, or any number. Since there are too many answers, 0 ÷ 0 is indeterminate.

PRINCIPLE 12 When you multiply or divide a number by 1, the answer is the number you started with.

The following models explain this principle for multiplication and for division.

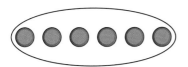

1 × 6 is 1 group of 6,
or 6 items altogether.

6 ÷ 1 is 6 items in 1 group,
or 6 items in each (of 1) group.

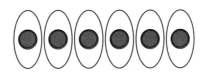

6 × 1 is 6 groups of 1,
or 6 items altogether.

6 ÷ 1 is 6 items in groups of 1,
or 6 groups altogether.

Using the Principles to Learn the Facts

Students who become comfortable with the principles listed on **page 27** will recognize which strategies might be useful in which situations to relate an unknown fact to a known one.

MULTIPLICATION AND DIVISION PRINCIPLE	EXAMPLE
By knowing Principle 1 (multiplication and division are inverse operations), students do not need to separately memorize the division facts, as they can use a related multiplication fact instead.	Since 4 × 6 = 24, 24 ÷ 4 = 6 and 24 ÷ 6 = 4.
By knowing Principle 2 (the commutative property), students only need to learn half of the multiplication facts.	If students know 6 × 5 = 30, then they know 5 × 6 = 30.
By knowing Principles 9 and 12, students do not need to separately memorize facts involving 1 or 0.	■ × 1 = ■ 1 × ■ = ■ ■ ÷ 1 = ■ ■ ÷ ■ = 1 no matter what number ■ is (except 0) ■ × 0 = 0 0 × ■ = 0 0 ÷ ■ = 0 no matter what number ■ is

Multiplication and Division Procedures

Multiplying and Dividing with Multiples of Powers of 10

Both estimation and calculation of multi-digit products and quotients are based on students knowing the multiplication and division facts, and knowing how to multiply and divide with multiples of 10, 100, 1000, etc. The following models can be used to teach and explain these concepts.

To provide a fun opportunity for students to apply different principles and strategies and to use their own strategies (BIWNO 2) to learn multiplication facts, they can play this game in pairs.

Each player uses a different colour counter and the game board shown below, which consists of a product chart and a factor strip.

Player A puts a counter on a number on the factor strip.

Player B puts a counter on the strip (on the same or a different number) and then multiplies the two numbers, putting a counter on the product in the product chart.

Player A moves his or her counter on the strip to another number, multiplies the two numbers, and puts a counter on the product.

Play continues until someone has four counters in a row on the product chart.

As students play, ask questions like: *Why did you move your counter to 9?* [e.g., I needed to cover 27 and I knew I could use the multiplication fact 3 × 9 = 27.]

1	2	3	4	5	6
7	8	9	10	12	14
15	16	18	20	21	24
25	27	28	30	32	35
36	40	42	45	48	49
54	56	63	64	72	81

1	2	3	4	5	6	7	8	9

Multiplying Using Place Value

Multiplying 1-digit by 2-digit Numbers: 5 × 30

Step 1 Model 5 × 30 as 5 groups of 3 tens.

$$5 \times 30 = 5 \times 3 \text{ tens}$$
$$= 15 \text{ tens}$$

Step 2 Regroup 15 tens by trading 10 tens for 1 hundred.

$$5 \times 3 \text{ tens} = 15 \text{ tens}$$
$$= 1 \text{ hundred, } 5 \text{ tens}$$
$$= 150$$

This model can be extended to 5 × 300 using base ten hundred blocks (flats) for hundreds, and to 5 × 3000 using base ten thousand blocks (large cubes) for thousands.

Multiplying 2-digit Numbers: 20 × 30

Step 1 Model 20 × 30 as 20 groups of 3 tens.

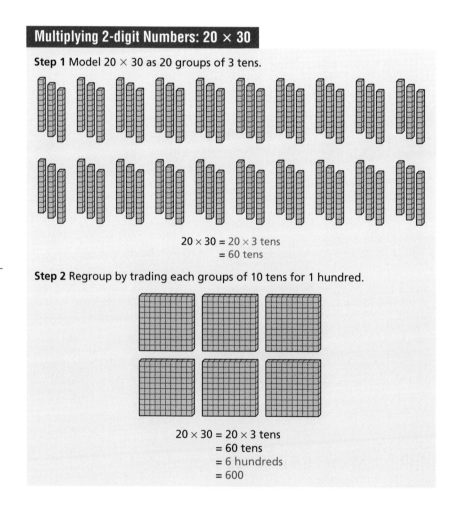

$$20 \times 30 = 20 \times 3 \text{ tens}$$
$$= 60 \text{ tens}$$

Step 2 Regroup by trading each groups of 10 tens for 1 hundred.

$$20 \times 30 = 20 \times 3 \text{ tens}$$
$$= 60 \text{ tens}$$
$$= 6 \text{ hundreds}$$
$$= 600$$

Teaching Idea | **2.22**

Ask students to look at the array of 6 hundred blocks shown in Step 2. Push them together to form a 20 by 30 rectangle.

To connect multiplication and area (BIWNO 3), ask: *How does this rectangle show 20 groups of 30?* [There are 20 rows of 30 ones.] *How does this help us see why 20 × 30 = 2 × 3 hundreds, or 600?* [20 rows of 3 ones = 2 rows of 3 hundreds, which is 600]

Note that students usually find calculations like 20 × 30, when both factors are multiples of 10, more difficult than when only one factor is a multiple of 10, such as 2 × 30 or 20 × 3.

Dividing Using Place Value

Dividing 320 ÷ 8 Using Base Ten Blocks

Step 1 Model 320 as 32 tens.

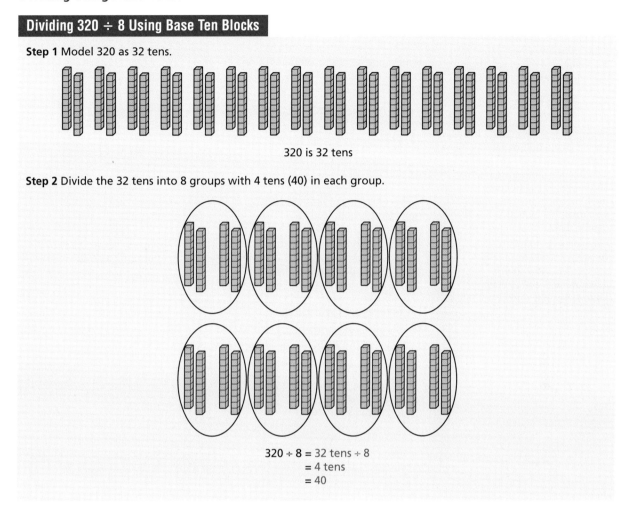

320 is 32 tens

Step 2 Divide the 32 tens into 8 groups with 4 tens (40) in each group.

$$320 \div 8 = 32 \text{ tens} \div 8$$
$$= 4 \text{ tens}$$
$$= 40$$

Note that this model can be extended to 3200 ÷ 8, using base ten hundred blocks (flats), and to 32 000 ÷ 8, using base ten thousand blocks (large cubes) for thousands.

Dividing 400 ÷ 80 Symbolically

$400 \div 80$
$= 40 \text{ tens} \div 8 \text{ tens}$ (Express each number using the same place value.)
$= 5$ (Divide 40 ÷ 8.)

40 tens ÷ 8 tens = 5 makes sense because 40 items divided into groups of 8 items is 5 groups.

Estimating Products and Quotients

A number of factors come into play when making decisions about estimating, including the context and the numbers and operations involved. Students should observe that estimates involving multiplication and division with greater values often tend to be further from the actual values than is the case when estimating with addition and subtraction. Students should pay particular attention to whether the estimated product or quotient might be too low or too high.

Teaching Idea | **2.23**

Focus on BIWNO 4 by helping students observe what happens when $346 \div 6$ is estimated each way:

$346 \div 6 \rightarrow 360 \div 6$ [about 60]

$346 \div 6 \rightarrow 300 \div 5$ [about 60]

$346 \div 6 \rightarrow 420 \div 7$ [about 60]

Discuss why each change in the dividend and divisor makes sense. [e.g., changing 6 to 5 while changing 346 to 300 makes sense because both numbers are decreased, and thus the fact $30 \div 5 = 6$ can be used]

Encourage students to show flexibility in estimation by asking: *How might you estimate 789 ÷ 8 two ways?* [$700 \div 7 = 100$, $810 \div 9 = 90$]

Estimating Strategies

To estimate products and quotients, students might use strategies such as:

ESTIMATION STRATEGY	EXAMPLES
Round one or both numbers to the nearest multiple of 10, 100, 1000, ...	25×52 is about $25 \times 50 = 1250$ 39×31 is about $40 \times 30 = 1200$ 642×32 is about $600 \div 30 = 20$
Round numbers such that familiar multiplication and division facts can be used.	$574 \div 9$ is about $560 \div 8 = 70$ $574 \div 9$ is about $540 \div 9 = 60$
When multiplying, round one factor up and the other factor down.	65×15 is about $60 \times 20 = 1200$
When dividing, round both numbers up or both numbers down.	$337 \div 8$ is about $360 \div 9 = 40$ $337 \div 8$ is about $280 \div 7 = 40$
Round numbers to the nearest multiple of 10, 100, 1000, ... or 25 to be able to multiply or divide by 25.	$389 \div 27$ is about $400 \div 25 = 16$ $612 \div 27$ is about $600 \div 25 = 24$

The Effects of Rounding

When multiplying, rounding one factor has a different effect than rounding the other. In the following example, rounding the 8 to 10 has a greater effect on the estimated product than rounding the 68 to 70, even though it is an increase of 2 for each. This is because two extra 68s in 68×10 is more than 8 extra 2s in 70×8.

MULTIPLYING EXACTLY	ROUNDING THE SECOND FACTOR UP	ROUNDING THE FIRST FACTOR UP
$68 \times 8 = 544$	$68 \times 10 = 680$ 680 is 136 greater than 544.	$70 \times 8 = 560$ 560 is only 16 greater than 544.

When dividing, rounding the dividend has a different effect than rounding the divisor. The following example uses the sharing model of division to explain why rounding the dividend up increases the estimated quotient, but rounding the divisor up decreases the estimated quotient.

DIVIDING EXACTLY	ROUNDING THE DIVIDEND UP	ROUNDING THE DIVISOR UP
$450 \div 7 \approx 64.3$ 450 items shared among 7 people is 64 items each, with some left over.	$450 \div 7$ is about $490 \div 7 = 70$ The estimated quotient, 70, is greater than the exact quotient, 64.3, since there are more items to share among the same number of people.	$450 \div 7$ is about $450 \div 9 = 50$ The estimated quotient, 50, is less than the exact quotient, 64.3, since there are more people sharing the same number of items.

Teaching Idea | **2.24**

Allow students to invent their own multiplication strategies before showing them a strategy you prefer (BIWNO 2).

After students have shared their strategies, ask: *What did you like best about her strategy? Why did your strategy and her strategy result in the same product?*

Varied Approaches for Multiplication and Division

Importance of Invented Approaches

There are many good reasons for students to be exposed to multiple algorithms for multiplication and division, and for them to invent their own strategies and algorithms.

- One algorithm might make more sense to a student than another.
- One algorithm might work better for a particular set of numbers.
- Some algorithms lend themselves to mental computation.

- A student may get help at home from a parent who uses a very different algorithm than what has been taught at school. It is helpful if students are open to both.
- Students who have a repertoire of algorithms to choose from can use one algorithm to perform a calculation, and a different one to check it.
- Some algorithms are actually procedures that a student might "invent." Something a student creates himself or herself is almost always more meaningful to him or her.

Alternative Multiplication Algorithms

For Algorithm 1 below (often considered the "traditional" algorithm) and many others, it is appropriate to initially model the algorithm with manipulatives. Normally, a written record is not necessary in the early stages of algorithm use, but is important later on, with each step of the algorithm matched to a physical action with the manipulatives.

A number of these algorithms translate well into mental algorithms.

Algorithm 1 The traditional multiplication algorithm is built on the principle that you can multiply in parts. To multiply 5×423, the number 423 is broken up into $400 + 20 + 3$, and each part is multiplied by 5. In this version of the traditional algorithm, the parts are calculated starting with the smaller values, and regrouping is done as you go.

No matter what algorithm is used, it is important that students understand and are able to explain what they are doing and why.

Algorithm 1 for Multiplying 5 × 423

Step 1 Model 423 as 5 groups of 4 hundreds, 2 tens, and 3 ones. (A place value mat is optional.)

Step 2 Combine the ones. Trade 10 ones for 1 ten. Record the ones that are left.

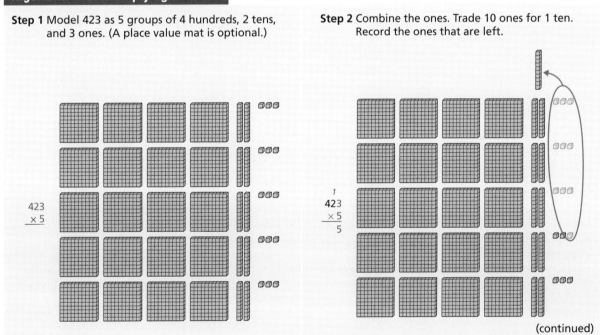

(continued)

Algorithm 1 for Multiplying 5 × 423 (continued)

Step 3 Combine the tens. Trade 10 tens for 1 hundred. Record the tens that are left.

Step 4 Record the hundreds.

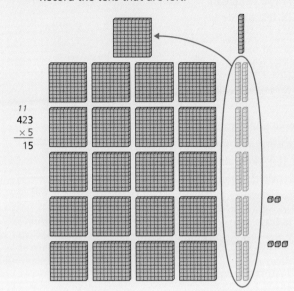

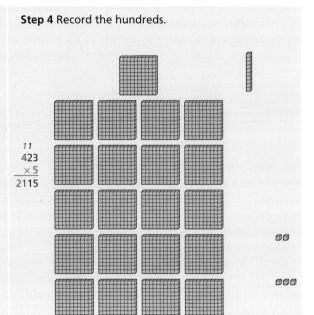

Note that in Algorithm 1, the partial products are not recorded because you regroup as you go, instead of at the end as in Algorithm 2.

Algorithm 2 This algorithm is sometimes called the partial product algorithm because all partial products are recorded, and then they are added up at the end. The multiplying can be done either beginning with the greater place values (Algorithm 2A) or beginning with the lower place values (Algorithm 2B).

ALGORITHM 2A	ALGORITHM 2B
423	423
× 5	× 5
2000 (5 × 400)	15 (5 × 3)
100 (5 × 20)	100 (5 × 20)
+ 15 (5 × 3)	+2000 (5 × 400)
2115 (Add the partial products.)	2115 (Add the partial products.)

MODELLING ALGORITHM 2

Base Ten Blocks

- 2A can be modelled using base ten blocks beginning with 5 groups of blocks, each with 4 hundred blocks, 2 ten blocks, and 3 one blocks (as in Step 1 of Algorithm 1).
- The next step is to combine the hundred blocks, then the ten blocks, and finally the one blocks (forming the three partial products in the recorded algorithm).
- The last step is any final regrouping of the blocks.

Base Ten Blocks

- 2B can be modelled using base ten blocks beginning with 5 groups of blocks, each with 4 hundred blocks, 2 ten blocks, and 3 one blocks (as in Step 1 of Algorithm 1).
- The next step is to combine the one blocks, then the ten blocks, and finally the hundred blocks (forming the three partial products).
- The last step is any final regrouping of the blocks.

Area Model

Algorithms 2A and 2B can also be modelled using the same area model, which is based on the area meaning of multiplication; that is, the area of a rectangle is the product of its length and width.

36 BIG IDEAS for Teaching Mathematics, Grades 4 to 8

Here is a pictorial area model of the multiplication:

5 × 423

Length is 423 or 400 + 20 + 3.

5 × 20 = 100

| Width is 5. | 5 | 5 × 400 = 2000 | 20 | 3 |

The area is 2000 + 100 + 15 = 2115.

5 × 3 = 15

Here are both symbolic models:

2A	2B
423	423
× 5	× 5
2000	15
100	100
+ 15	+2000
2115	2115

Multiplying Two 2-digit Numbers

To multiply two 2-digit numbers, students can apply and combine the various models already presented.

For example, to multiply 32 × 43, they might create an area model using base ten blocks.

A BASE TEN BLOCK AREA MODEL FOR MULTIPLYING 2-DIGIT NUMBERS

A rectangle of blocks that is 32 by 43

Each of the four parts of the rectangle is a partial product.

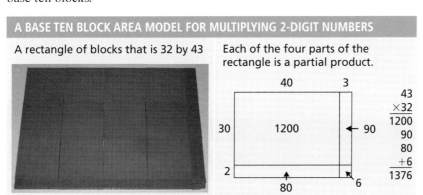

	40	3	
30	1200	← 90	43
			×32
			1200
			90
2			80
	80	6	+6
			1376

To multiply 53 × 34, they might use the lattice algorithm.

THE LATTICE ALGORITHM FOR MULTIPLYING 2-DIGIT NUMBERS (53 × 34)

Step 1 To multiply 53 × 34, draw a 2 by 2 grid, with each square divided diagonally as shown.

Record the factors outside the squares.

Record the digits of one factor across the top (5 and 3) and the digits of the other factor (3 and 4) down the right.

Step 2 For each square of the grid, multiply the two outside corresponding values.

Record the tens digit (if there is one) of the product above the diagonal and the ones digit below:

5 × 3 = 15   3 × 3 = 9
5 × 4 = 20   3 × 4 = 12

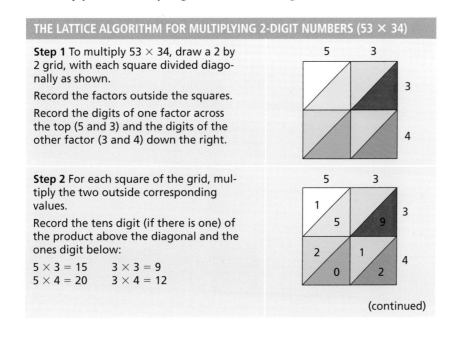

(continued)

> **Teaching Idea | 2.25**
>
> Students might find this algorithm interesting.
>
> 35
> × 28
> 640 [20 × 30 + 8 × 5]
> + 340 [8 × 30 + 20 × 5]
> 980
>
> To emphasize the variety of algorithms (BIWNO 1), ask: *How is this strategy like yours? How is it different?*

> *The lattice algorithm works because each digit is automatically placed in the correct place value column. Although it is fairly simple for students to follow, it does not bring meaning to the steps of the operation.*

Step 3 Add the inside values along each diagonal from the right down to the left. Regroup if necessary to the next diagonal:
- record 2
- 9 + 1 + 0 = 10, so record 0 and regroup 1
- 1 + 5 + 2 = 8, so record 8
- record 1

The values down the left and across the bottom (1802) are the final product.

53 × 34 = 1802

Alternative Division Algorithms

A number of these division algorithms translate well into mental algorithms.

As with multiplication, it is appropriate to initially model many of the division algorithms with manipulatives. A written record is not necessary in the early stages of algorithm use, but is important later on, with each step of the algorithm matched to a physical action with the manipulatives.

Algorithm 1 This algorithm is built on the numeration principle that a number has many different "forms," and on the division principle that you can divide in parts. This traditional algorithm is best demonstrated using the sharing meaning of division (346 items shared among 3 people) and a concrete base ten model, as shown below.

Algorithm 1 is best demonstrated using the sharing meaning of division.

*Algorithm 2 is best demonstrated using the grouping meaning (as shown on **page 40**).*

Algorithm 1 for Dividing 346 ÷ 3

Step 1 Model 346 with 3 hundred blocks, 4 ten blocks, and 6 one blocks. Draw 3 boxes to represent each "share."

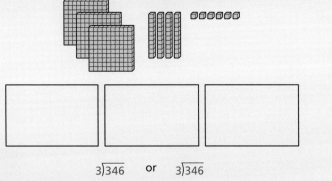

3)346   or   3)346

Two alternative written forms of the algorithm are presented.

Step 2 Share the 3 hundred blocks. Each shape gets 1 hundred block, so record 100 or 1 (hundred). There are 4 ten blocks and 6 one blocks left, so record 46.

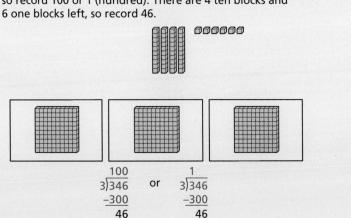

$$\begin{array}{r} 100 \\ 3\overline{)346} \\ -300 \\ \hline 46 \end{array} \quad \text{or} \quad \begin{array}{r} 1 \\ 3\overline{)346} \\ -300 \\ \hline 46 \end{array}$$

Step 3 Share the 4 ten blocks. Each shape gets 1 ten block, so record 10 or 1 (ten). Trade the leftover ten block for 10 one blocks (to be shared in Step 4). There are 16 one blocks left, so record 16.

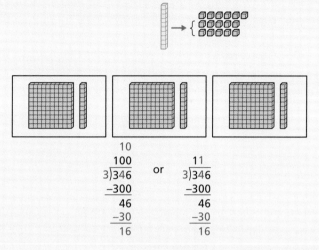

$$
\begin{array}{r}
10 \\
100 \\
3\overline{)346} \\
-300 \\
\hline
46 \\
-30 \\
\hline
16
\end{array}
\quad \text{or} \quad
\begin{array}{r}
11 \\
3\overline{)346} \\
-300 \\
\hline
46 \\
-30 \\
\hline
16
\end{array}
$$

Step 4 Share the 16 one blocks. Each share gets 5 one blocks, so record 5. One block is left over as a remainder.

R 1 🔲

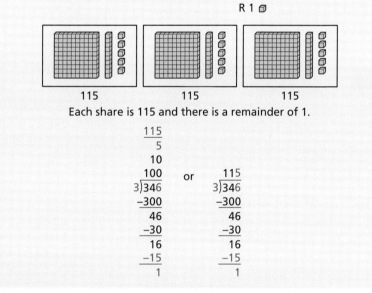

115   115   115

Each share is 115 and there is a remainder of 1.

$$
\begin{array}{r}
115 \\
5 \\
10 \\
100 \\
3\overline{)346} \\
-300 \\
\hline
46 \\
-30 \\
\hline
16 \\
-15 \\
\hline
1
\end{array}
\quad \text{or} \quad
\begin{array}{r}
115 \\
3\overline{)346} \\
-300 \\
\hline
46 \\
-30 \\
\hline
16 \\
-15 \\
\hline
1
\end{array}
$$

Algorithm 2 As with Algorithm 1, this algorithm is based on the principle that you can divide in parts by splitting the dividend into parts. This algorithm makes most sense to model using the equal groups meaning of division (How many groups of 3 are in 346?), as shown.

The algorithm on the right shows how you can underestimate the number of groups and still successfully complete the algorithm, although you end up taking more steps to do it. (This is not the case with Algorithm 1.)

Teaching Idea | **2.26**

Introduce Egyptian multiplication to reinforce BIWNO 1. With this algorithm, students only need to double and add. For 36 × 53,

Start two columns, one beginning with 1 and the other with 53.

Keep doubling until the number in the 1 column is not yet greater than 36.

Find numbers in the first column that add to 36 (4 + 32).

Add the corresponding numbers in the second column to get the product (212 + 1696 = 1908).

1	53
2	106
4	**212**
8	424
16	848
32	**1696**

Have students use this method and then bring out the connection between multiplication and division and BIWNO 2 by asking: *How could you use this idea to develop a procedure for dividing 528 ÷ 8?*

[1	8
2	16
4	32
8	64
16	128
32	256
64	512

512 + 16 = 528, so
528 ÷ 8 = 64 + 2 = 66]

- How many times would you have to write your first name to have at least 1000 letters?

- How long would it take you to walk 1000 paces (if you timed how long it took you to walk 10 paces)?

- What is the "average" distance you can throw a ball over five throws?

Emphasize the value of alternative strategies (BIWNO 1), by asking: *How did you solve that problem? How else could you have solved it?*

Algorithm 2 for Dividing 346 ÷ 3

DIVIDING IN PARTS	UNDERESTIMATING EACH TIME

DIVIDING IN PARTS

$$
\begin{array}{r}
3\overline{)346} \\
-300 \quad\quad 100 \;\; (100 \text{ groups of } 3)\\
\overline{46} \\
-30 \quad\quad 10 \;\; (10 \text{ groups of } 3)\\
\overline{16} \\
-15 \quad\quad \underline{5} \;\; (5 \text{ groups of } 3)\\
\overline{1} \quad\quad 115 \text{ R } 1
\end{array}
$$

UNDERESTIMATING EACH TIME

$$
\begin{array}{r}
3\overline{)346} \\
-150 \quad\quad 50 \;\; (50 \text{ groups of } 3)\\
\overline{196} \\
-60 \quad\quad 20 \;\; (20 \text{ groups of } 3)\\
\overline{136} \\
-90 \quad\quad 30 \;\; (30 \text{ groups of } 3)\\
\overline{46} \\
-30 \quad\quad 10 \;\; (10 \text{ groups of } 3)\\
\overline{16} \\
-15 \quad\quad \underline{5} \;\; (5 \text{ groups of } 3)\\
\overline{1} \quad\quad 115 \text{ R } 1
\end{array}
$$

Algorithm 3 This algorithm is based on breaking up the dividend into "comfortable" parts, that is, into parts, or numbers, that are easy to divide. The underlying principle is that you can divide by splitting the dividend into parts (Principle 7). This makes sense if you visualize the sharing model of division. For example, for the first example shown below, the steps are like sharing 400 items among 3 people, by first sharing 300 among the 3 people, then 90 among the 3 people, and then the last 10 among the 3 people.

Variations of Algorithm 3 for 400 ÷ 3

DIVIDING 400 AS 300 + 90 + 10	DIVIDING 400 AS 150 + 240 + 10	DIVIDING 400 AS 390 + 10

$$
\begin{array}{l}
\quad\quad\quad\quad 133 \text{ R } 1 \\
100 + 30 + 3 \text{ R } 1 \\
\overline{3)300 + 90 + 10}
\end{array}
$$

$$
\begin{array}{l}
\quad\quad\quad\quad 133 \text{ R } 1 \\
50 + 80 + 3 \text{ R } 1 \\
\overline{3)150 + 240 + 10}
\end{array}
$$

$$
\begin{array}{l}
\quad\quad\quad 133 \text{ R } 1 \\
130 + 3 \text{ R } 1 \\
\overline{3)390 + 10}
\end{array}
$$

Treatment of Remainders

When dividing whole numbers, there are sometimes leftover amounts. These form the "remainder." Students need to learn to make sense of the remainder conceptually, as well as how to account for the remainder symbolically when using a division algorithm.

Using Context to Interpret Remainders

Remainder as a Whole Number

34 marbles shared among 3 children:

$34 \div 3 = 11 \text{ R } 1$

It does not make sense to divide the remaining 1 marble further. So, each child gets 11 marbles, and there is 1 marble remaining.

Remainder as a Fraction

34 hours shared among 3 workers:

$$
\begin{array}{r}
11 \\
3\overline{)34} \\
-33 \\
\overline{1}
\end{array}
$$

The remaining 1 hour can be divided among the 3 workers so each gets $\frac{1}{3}$ of an hour. So, each worker gets $11\frac{1}{3}$ hours.

Remainder as a Decimal

$34.00 shared among 3 people:

```
        11.33
   3)34.00
     -33
      10
      -9
       1
```

It does not make sense to divide the remaining 1 cent further, since there is no such thing as $11.333. So, each person gets $11.33 and there is $0.01 (1 cent) remaining.

Rounding a Remainder Up

There are 34 children at 3 per car. How many cars are needed?

$34 \div 3 = 11 \text{ R } 1$

An additional car is needed for the remaining 1 child. (It does not make sense to leave 1 child behind, and 11.33 or $11\frac{1}{3}$ cars does not make sense.) So, 12 cars will be needed for 34 children.

Context is what determines how a remainder should be treated.

Mental Calculation

Mental calculation suits certain number combinations better than others. As well, certain algorithms suit mental computation better than others. For example,

USING DIVISION ALGORITHM 3	USING MULTIPLICATION ALGORITHM 2
Students might calculate $512 \div 3$ mentally using division algorithm 3 by thinking of 512 in "comfortable" parts ($300 + 210 + 2$) that can be divided mentally by 3. $\begin{array}{r} 170 \text{ R } 2 \\ 100 + 70 + 0 \text{ R } 2 \\ \hline 3)300 + 210 + 2 \end{array}$	Some people find multiplication algorithm 2 suitable for mental calculation, particularly if one factor is a 1-digit number and especially if there is limited regrouping when the partial products are added, although this depends entirely on the ability of individual students. They need to keep each partial product "in their head" as they multiply and then add them up mentally. $\begin{aligned} 2 \times 341 &= 2 \times 300 + 2 \times 40 + 2 \times 1 \\ &= 600 + 80 + 2 \\ &= 682 \end{aligned}$

Other possible mental approaches, as described below, can be generalized to similar situations:

- 47×28 as 4 tens $\times$ 2 tens + 7 ones $\times$ 8 ones + $(7 \times 2 + 4 \times 8)$ tens. This is $800 + 56 + 460$. You can calculate $856 + 460$ by adding 500 to 856 and then subtracting 40 to get 1316.
- 54×50 is half of 54×100. So if $54 \times 100 = 5400$, then $54 \times 50 = 5400 \div 2 = 2700$.
- 448×25 is one-fourth of 448×100. So, if $448 \times 100 = 44\,800$, then $448 \times 25 = 44\,800 \div 4 = 11\,200$.
- $650 \div 5$ is twice as much as $650 \div 10$. So, if $650 \div 10 = 65$, then $650 \div 5 = 65 \times 2 = 130$.

Whether a calculation can be done mentally depends on the numbers involved, the algorithm used, and the ability of the student.

Fractions

In the first three grades, students are introduced to fractions primarily, but not exclusively, as part of a region, for example, a half of a shape. In Grades 4 to 6, students expand their understanding of fractions by working with a broader variety of fraction meanings and representations and focusing on equivalence and comparison. In Grades 6 to 8, students expand their understanding to include the meaning of fractions as implied division, and they develop an understanding of fraction operations.

BIG IDEAS FOR FRACTIONS

Each teaching idea in this section of the chapter will indicate which Big Idea(s) for Fractions (BIF) can be emphasized.

1. Fractions can represent parts of regions, parts of sets, parts of measures, division, or ratios. These meanings are equivalent, e.g., $\frac{1}{3}$ of a region is 1 whole divided into 3 equal parts.

2. A fraction is not meaningful without knowing what the whole is.

3. Renaming fractions is often the key to comparing them or computing with them. Every fraction can be renamed in an infinite number of ways.

4. There are multiple models and /or procedures for comparing and computing with fractions, just as with whole numbers.

5. Operations with fractions have the same meanings as operations with whole numbers, even though the algorithms differ.

Fractions in the Child's World

Even though a fraction has two numbers, it is one idea—the relationship between the two numbers.

A fraction is a number that describes a relationship between a part (represented by the numerator) and a whole (represented by the denominator). Although students see two numbers when they look at a fraction, they have to think of one idea—the relationship between the two numbers. As a result, fractions are a challenge for many students and even adults.

Some basic fraction terms used in the teaching of fractions are listed below:

FRACTION DEFINITIONS

$$\underline{3} \rightarrow \text{numerator}$$
$$4 \rightarrow \text{denominator}$$

- the denominator tells how many equal parts the whole is divided into
- the numerator tells how many there are of those equal parts
- a unit fraction has a numerator of 1, e.g., $\frac{1}{3}$
- a proper fraction, e.g., $\frac{4}{5}$, is less than 1
- an improper fraction, e.g., $\frac{7}{5}$, has a numerator equal to or greater than the denominator
- a mixed number, e.g., $5\frac{7}{8}$, is a whole number and a proper fraction

The first fraction that students meet is usually $\frac{1}{2}$. Generally, they work next with other unit fractions, such as $\frac{1}{4}$ and $\frac{1}{3}$. Once students have a firm grasp of the commonly used unit fractions, they typically extend their work to other proper fractions, such as $\frac{2}{3}$, $\frac{3}{4}$, and $\frac{5}{8}$. In later grades, they meet improper fractions like $\frac{3}{2}$, and mixed numbers such as $1\frac{1}{2}$ and $2\frac{3}{5}$. Eventually, students work with fraction operations, but usually not until middle school.

Fraction Meanings

Fractions are used to represent parts of a region, parts of a measure, and parts of a set (or group); they can also represent division and ratios. Initially, students experience fractions as parts of a region and parts of a set. Later they are introduced to the other meanings of fractions.

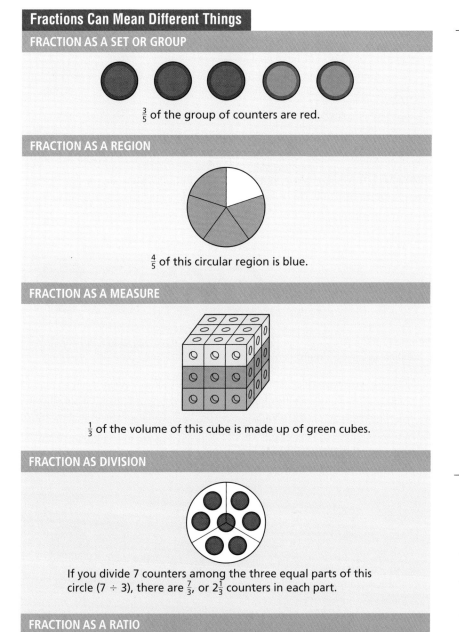

Fractions Can Mean Different Things

FRACTION AS A SET OR GROUP

$\frac{3}{5}$ of the group of counters are red.

FRACTION AS A REGION

$\frac{4}{5}$ of this circular region is blue.

FRACTION AS A MEASURE

$\frac{1}{3}$ of the volume of this cube is made up of green cubes.

FRACTION AS DIVISION

If you divide 7 counters among the three equal parts of this circle (7 ÷ 3), there are $\frac{7}{3}$, or $2\frac{1}{3}$ counters in each part.

FRACTION AS A RATIO

The ratio of red hexagons to total hexagons is 2 : 5, or $\frac{2}{5}$ of this set of hexagons is red.

Teaching Idea | **2.28**

Ask students to draw as many pictures as they can of $\frac{2}{5}$ to see what fraction meanings they are most comfortable with.

Focus on multiple fraction meanings (BIF 1) by asking: *How are Aaron's and your picture of $\frac{2}{5}$ the same?* [e.g., both show 2 parts out of 5] *different?* [e.g., mine is 2 out of 5 circles and Aaron's is 2 equal parts of a circle] *If they are different, how can they both show $\frac{2}{5}$?* [Both times, there is a whole and 2 out of 5 parts of the whole are being used.]

Teaching Idea | **2.29**

Help students connect fraction meanings (BIF 1) such as $\frac{5}{6}$ is 5 parts of a region divided into 6 equal parts and $\frac{5}{6}$ is 5 ÷ 6 by asking them how they would share 5 pizzas among 6 people?

To emphasize BIF 1, ask: *How would knowing $\frac{1}{6} = 1 ÷ 6$ help you solve this problem?* [I'd divide each pizza into 6 pieces. Each person gets $\frac{1}{6}$ of each pizza, so each gets $\frac{5}{6}$ altogether.]

Fraction Principles

Some of the important principles that students in Grades 4 to 8 must learn about fractions are listed below.

FRACTION PRINCIPLES

1. A fraction has a numerator and a denominator.
2. You have to know what the whole is to know what the fraction is.
3. The equal parts into which the whole is divided are equal but do not have to be identical. For parts of a set, this means that the members of the set do not have to be identical.
4. Fraction parts do not need to be adjacent.
5. If the numerator and denominator of a fraction are equal, the fraction represents one whole, or 1. This means all whole numbers can be represented as fractions.
6. Fractions have more than one name.
7. Fractions with numerators greater than their denominators are greater than 1.

PRINCIPLE 1 A fraction has a numerator and a denominator.

The denominator tells the total number of equal parts the whole is divided into, and the numerator tells the number of those parts there are.

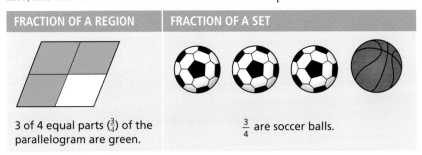

FRACTION OF A REGION	FRACTION OF A SET
3 of 4 equal parts ($\frac{3}{4}$) of the parallelogram are green.	$\frac{3}{4}$ are soccer balls.

PRINCIPLE 2 You have to know what the whole is to know what the fraction is.

For example, if we say the blue rectangle below is $\frac{1}{4}$, it is only in relation to the rectangle on the left. In relation to the rectangle on the right, the blue rectangle is $\frac{1}{6}$. In the example of the family below, the fraction used to describe the girl depends on what is considered the whole, whether it is the children or the whole family.

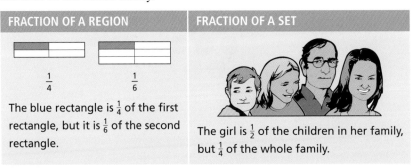

FRACTION OF A REGION	FRACTION OF A SET
$\frac{1}{4}$ $\qquad$ $\frac{1}{6}$	
The blue rectangle is $\frac{1}{4}$ of the first rectangle, but it is $\frac{1}{6}$ of the second rectangle.	The girl is $\frac{1}{2}$ of the children in her family, but $\frac{1}{4}$ of the whole family.

PRINCIPLE 3 The equal parts into which the whole is divided are equal but do not have to be identical. For parts of a set, this means that the members of the set do not have to be identical.

The rectangle at the top left of the next page is divided into four equal parts, each a fourth, but notice that not all of the fourths are congruent; for example, the yellow fourth is wide and short and the green fourth is narrower and taller. However, they are still equal parts because each fourth has the same area.

Teaching Idea | **2.30**

We often show students a whole and have them identify a fraction part. Reverse this by drawing a rectangle on grid paper, telling the students that it represents, say, $\frac{1}{4}$ of a whole, and then asking them to draw what the whole might look like.

To emphasize BIF 2, ask: *How did you use the denominator and numerator to help you figure out what the whole might look like?*
[I knew the rectangle was 1 part out of 4, so I drew another rectangle that was 4 times as big in area.]

Teaching Idea | **2.31**

To focus on BIF 2, begin with whole numbers and talk about why 100 can sometimes be 4 of something, but sometimes 10 of something (e.g., 100 cents = 4 quarters or 10 dimes). Apply this to fractions by showing two egg cartons, one full of eggs and the other half full and ask: *When would this be 1$\frac{1}{2}$ full?* [If the whole is 1 carton] *$\frac{3}{4}$ full?* [If the whole is 2 cartons] *What other number or fraction could you use?* [e.g., 3 if the whole is 6 eggs; $\frac{1}{2}$ if the whole is 3 cartons]

This rectangle is divided into non-congruent fourths.

The green part and the yellow part of the rectangle may not be congruent, but they are equal in area (as each is made up of the same two congruent right triangles). That means that each is a fourth of the rectangle.

In the group of shapes below, $\frac{3}{5}$ are red because, even though each fifth is a different shape and size, the whole is considered the set of 5 shapes, so each shape is 1 fifth.

Each member of this set of shapes is different.

$\frac{3}{5}$ of this set of shapes is red.

PRINCIPLE 4 Fraction parts do not need to be adjacent.

In this four-coloured rectangle, the two sections that make up the yellow fourth are not adjacent; neither are the two sections that make up the blue fourth. It is important for teachers to attend to this. In some cases, teachers so rarely show fractions with non-adjacent parts that students are uncomfortable when they encounter them.

This rectangle is divided into fourths by colour.

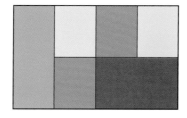

One fourth of the rectangle is blue and one fourth is yellow, even though the sections that make up the blue fourth and the yellow fourth are not adjacent.

PRINCIPLE 5 If the numerator and denominator of a fraction are equal, the fraction represents one whole, or 1. This means all whole numbers can be represented as fractions.

This principle holds true for $\frac{2}{2}$, $\frac{3}{3}$, $\frac{4}{4}$, $\frac{5}{5}$, $\frac{6}{6}$, ...

Teaching Idea | **2.32**

Ask students to use pattern blocks to show fractions such as $\frac{1}{6}$ or $\frac{2}{3}$. They will likely use the fraction of a region meaning, like this:

$\frac{1}{6}$ of the hexagon is green.

$\frac{2}{3}$ of the hexagon is blue.

Then ask them to show $\frac{1}{5}$ to force them to consider the fraction of a set meaning. For example,

$\frac{1}{5}$ of this set of blocks is yellow.

To emphasize BIF 1, ask: *Why is it easy to represent any fraction using the fraction of a set meaning?*
[You just need a number of items in the set equal to the denominator.]

FRACTION OF A REGION EXAMPLE	FRACTION OF A SET EXAMPLE

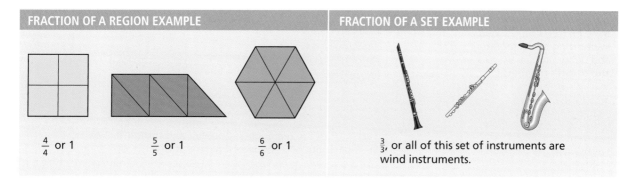

$\frac{4}{4}$ or 1 $\frac{5}{5}$ or 1 $\frac{6}{6}$ or 1

$\frac{3}{3}$, or all of this set of instruments are wind instruments.

PRINCIPLE 6 Fractions have more than one name.

In both examples below $\frac{1}{2}$ is another name for $\frac{2}{4}$, since the same part of the same whole is represented by both fractions.

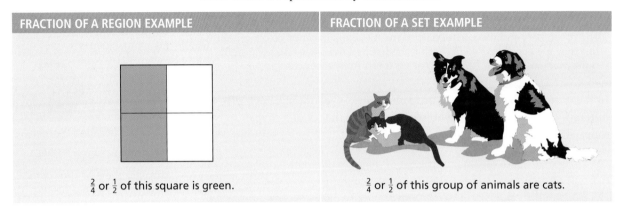

FRACTION OF A REGION EXAMPLE	FRACTION OF A SET EXAMPLE

$\frac{2}{4}$ or $\frac{1}{2}$ of this square is green. $\frac{2}{4}$ or $\frac{1}{2}$ of this group of animals are cats.

PRINCIPLE 7 Fractions with numerators greater than their denominators are greater than 1.

Some students assume something is wrong with improper fractions because of the term "improper," but this is, of course, not the case.

The pair of circles below could be described as $\frac{7}{4}$ red, or 7 fourths red if one circle is considered the whole. Since it only takes 4 fourths to make a whole (a circle), 7 fourths must be greater than a whole, or 1. The two egg cartons below could be described as $\frac{17}{12}$ full, or 17 twelfths full if one carton is considered the whole. Since one carton is 12 twelfths, 17 twelfths must be greater than a whole, or 1.

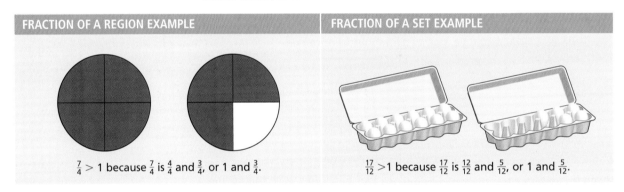

FRACTION OF A REGION EXAMPLE	FRACTION OF A SET EXAMPLE

$\frac{7}{4} > 1$ because $\frac{7}{4}$ is $\frac{4}{4}$ and $\frac{3}{4}$, or 1 and $\frac{3}{4}$. $\frac{17}{12} > 1$ because $\frac{17}{12}$ is $\frac{12}{12}$ and $\frac{5}{12}$, or 1 and $\frac{5}{12}$.

Equivalent Fractions

Teaching Idea | **2.33**

To emphasize BIF 3, ask: How do you know you can never "run out" of equivalent fractions for $\frac{2}{3}$? [I can just keep multiplying the numerator and denominators by greater and greater numbers.]

As described in Principle 6, fractions have more than one name. To find an equivalent name for a proper fraction, you can subdivide the existing parts to create more equal-size parts, which symbolically has the effect of multiplying the numerator and denominator by the same amount. For example,

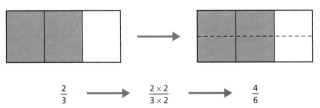

$$\frac{2}{3} \longrightarrow \frac{2 \times 2}{3 \times 2} \longrightarrow \frac{4}{6}$$

Each part of the rectangle is doubled and turns into 2 parts. So, 2 parts out of 3 becomes 4 parts out of 6.

One way to model subdividing fraction parts to create equivalent fractions is shown below.

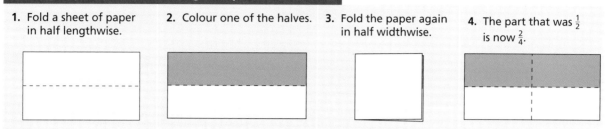

Folding Paper to Model Creating an Equivalent Fraction

1. Fold a sheet of paper in half lengthwise.

2. Colour one of the halves.

3. Fold the paper again in half widthwise.

4. The part that was $\frac{1}{2}$ is now $\frac{2}{4}$.

To create an equivalent name for a fraction, you can also combine parts, which symbolically has the same effect as dividing the numerator and denominator by the same amount. For example,

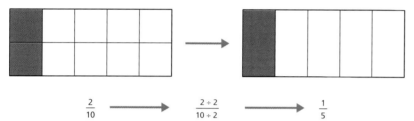

$$\frac{2}{10} \longrightarrow \frac{2 \div 2}{10 \div 2} \longrightarrow \frac{1}{5}$$

Each of the 2 parts of the rectangle is combined into 1 part.
So, 2 parts out of 10 becomes 1 part out of 5.

Mixed Numbers and Improper Fractions

One of the equivalences students need to learn about is the relationship between mixed numbers and their equivalent improper fractions. For example, $\frac{15}{4} = 3\frac{3}{4}$ since 15 fourths makes 3 wholes (4 fourths + 4 fourths + 4 fourths) and 3 fourths. Symbolically, this has the effect of the familiar procedure for changing an improper fraction to a mixed number, that is, dividing 15 by 4 to determine the number of wholes. The remainder is the number of fourths in the fraction part of the mixed number.

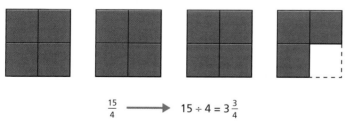

$$\frac{15}{4} \longrightarrow 15 \div 4 = 3\frac{3}{4}$$

Divide the 15 parts by 4 (because there are 4 parts per whole) to determine the number of wholes. The remainder becomes the fraction part of the mixed number.

Teaching Idea | **2.34**

Students can use number sense and apply BIF 3 by using all the digits from 1 to 9 to create a fraction equivalent to $\frac{1}{2}$. [e.g., $\frac{7692}{15384}$]

To focus students on the relationship between the numerators and denominators of equivalent fractions, in this case, that the denominator must be twice the numerator, ask: *How do you know the numerator will have four digits?* [$\frac{1}{2}$ is less than 1 so the numerator has to be less than the denominator; since there are 9 digits altogether, there must be 4 in the numerator and 5 in the denominator] *How do you know that the ones digit in the denominator will be even?* [e.g., the fraction equals $\frac{1}{2}$, so the denominator is a multiple of 2]

Students should not rush into the procedure of dividing the numerator by the denominator to change an improper fraction to a mixed number until the underlying meaning is clear.

Principles for Comparing Fractions

Some of the important principles that students must understand in order to compare fractions are listed below.

PRINCIPLES FOR COMPARING FRACTIONS

1. Fractions can be compared only if the whole is known in each case.
2. If two fractions have the same denominator, the one with the greater numerator is greater, e.g., $\frac{4}{5} > \frac{3}{5}$.
3. If two fractions have the same numerator, the one with the greater denominator is less, e.g., $\frac{3}{5} < \frac{3}{4}$.
4. Some fractions can be compared by relating them to benchmark numbers such as 0, 1, and $\frac{1}{2}$.
5. Fractions can be compared by renaming them with common denominators, or by renaming them with common numerators.
6. No matter what two different fractions are selected, there is always a fraction in between, e.g., $\frac{3}{6}$ is between $\frac{1}{3}$ and $\frac{2}{3}$.

Teaching Idea | 2.35

To connect fractions and whole numbers in terms of the importance of the unit or whole when comparing (BIF 2), ask: *Jana has three base ten blocks. Ian has four base ten blocks. Whose number is greater?* [You can't tell for sure until you know what blocks each has.] *I have $\frac{2}{3}$ of a container of ice cream left. Ian has $\frac{3}{4}$ of a container left. Who has more ice cream left?* [You can't tell for sure until you know what size each container is.]

PRINCIPLE 1 Fractions can be compared only if the whole is known in each case.

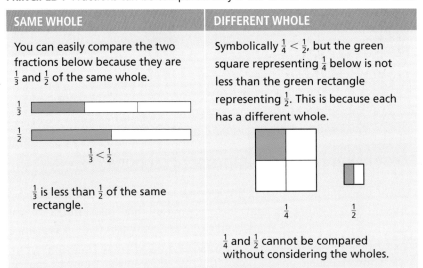

SAME WHOLE

You can easily compare the two fractions below because they are $\frac{1}{3}$ and $\frac{1}{2}$ of the same whole.

$$\frac{1}{3}$$

$$\frac{1}{2}$$

$$\frac{1}{3} < \frac{1}{2}$$

$\frac{1}{3}$ is less than $\frac{1}{2}$ of the same rectangle.

DIFFERENT WHOLE

Symbolically $\frac{1}{4} < \frac{1}{2}$, but the green square representing $\frac{1}{4}$ below is not less than the green rectangle representing $\frac{1}{2}$. This is because each has a different whole.

$$\frac{1}{4}$$

$$\frac{1}{2}$$

$\frac{1}{4}$ and $\frac{1}{2}$ cannot be compared without considering the wholes.

PRINCIPLE 2 If two fractions have the same denominator, the one with the greater numerator is greater.

This principle assumes that the fractions are fractions of the same whole. The denominator tells the total number of equal parts the whole is divided into, and the numerator tells the number of parts accounted for. If the denominators of two fractions are the same, then the parts are the same. So, the fraction with the greater number of parts accounted for is the greater fraction.

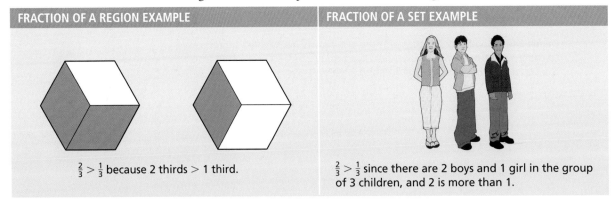

FRACTION OF A REGION EXAMPLE

$\frac{2}{3} > \frac{1}{3}$ because 2 thirds > 1 third.

FRACTION OF A SET EXAMPLE

$\frac{2}{3} > \frac{1}{3}$ since there are 2 boys and 1 girl in the group of 3 children, and 2 is more than 1.

PRINCIPLE 3 If two fractions have the same numerator, the one with the greater denominator is less.

This principle assumes that the fractions are fractions of the same whole. The denominator tells the total number of equal parts that the whole is divided into, and the numerator tells the number of parts accounted for. If the numerators are the same, then the number of parts accounted for is the same. But, if the denominators are different, then the fraction with the greater denominator is less, because the greater the number of parts, the smaller the parts.

FRACTION OF A REGION EXAMPLE	FRACTION OF A SET EXAMPLE
$\frac{2}{3} > \frac{2}{5}$ because 2 equal parts of a circle divided into 3 is greater than 2 equal parts of the same circle divided into 5.	$\frac{2}{4} < \frac{2}{3}$ since the girls make up less of the whole in the first group than in the second group.

PRINCIPLE 4 Some fractions can be compared by relating them to benchmark numbers such as 0, 1, and $\frac{1}{2}$.

If one fraction is greater than $\frac{1}{2}$ or 1 and the other is less than $\frac{1}{2}$ or 1, it is easy to compare them.

USING THE BENCHMARK $\frac{1}{2}$	USING THE BENCHMARK 1
$\frac{2}{5} < \frac{3}{4}$, since $\frac{2}{5} < \frac{1}{2}$ and $\frac{3}{4} > \frac{1}{2}$.	$\frac{7}{3} > \frac{3}{4}$, since $\frac{7}{3} > 1$ and $\frac{3}{4} < 1$.

PRINCIPLE 5 Fractions can be compared by renaming them with common denominators, or by renaming them with common numerators.

Here are two ways to compare $\frac{3}{8}$ and $\frac{4}{10}$. The first method is based on Principle 2, and the second method is based on Principle 3.

Renaming to Compare

WITH COMMON DENOMINATORS	WITH COMMON NUMERATOR
Comparing $\frac{3}{8}$ and $\frac{4}{10}$:	Comparing $\frac{3}{8}$ and $\frac{4}{10}$:
$\frac{3}{8} = \frac{30}{80}$ and $\frac{4}{10} = \frac{32}{80}$	$\frac{3}{8} = \frac{12}{32}$ and $\frac{4}{10} = \frac{12}{30}$
Since $\frac{30}{80} < \frac{32}{80}$, then $\frac{3}{8} < \frac{4}{10}$.	Since $\frac{12}{32} < \frac{12}{30}$, then $\frac{3}{8} < \frac{4}{10}$.

The strategy of using common numerators to compare fractions (that is, comparing the denominators of fractions if the numerators are the same) is often overlooked. In fact, students will often rename fractions, for instance, $\frac{2}{3}$ and $\frac{2}{5}$ as $\frac{10}{15}$ and $\frac{6}{15}$ in order to compare them, instead of simply comparing the denominators without doing any renaming at all.

Teaching Idea | **2.36**

Ask students to explain how they know $\frac{2}{3} > \frac{5}{8}$, in two different ways. [e.g.,

Using grids:

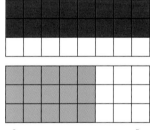

$\frac{2}{3}$ covers more area than $\frac{5}{8}$, so $\frac{2}{3} > \frac{5}{8}$.

Using equivalent fractions with a common numerator:
$\frac{2}{3} = \frac{10}{15}$ and $\frac{5}{8} = \frac{10}{16}$
Since $\frac{1}{15} > \frac{1}{16}$, then $\frac{10}{15} > \frac{10}{16}$, so $\frac{10}{15}$ or $\frac{2}{3}$ is greater.]

Point out the value of flexibility in approach and emphasize BIF 4 by asking: *Which strategy would you use to compare $\frac{2}{37}$ and $\frac{4}{73}$? Why?* [common numerators; finding the LCM of 2 and 4 is easy]

PRINCIPLE 6 No matter what two different fractions are selected, there is a fraction in between.

For instance, between $\frac{3}{5}$ and $\frac{4}{5}$ are an infinite number of fractions. You can rename $\frac{3}{5}$ and $\frac{4}{5}$ with greater denominators to find them, as shown below.

$\frac{3}{5} = \frac{6}{10}$ and $\frac{4}{5} = \frac{8}{10}$ so $\frac{7}{10}$ is between $\frac{3}{5}$ and $\frac{4}{5}$

$\frac{3}{5} = \frac{12}{20}$ and $\frac{4}{5} = \frac{16}{20}$ so $\frac{13}{20}, \frac{14}{20}$, and $\frac{15}{20}$ are between $\frac{3}{5}$ and $\frac{4}{5}$

$\frac{3}{5} = \frac{30}{50}$ and $\frac{4}{5} = \frac{40}{50}$ so $\frac{31}{50}, \frac{32}{50}, \dots , \frac{39}{50}$ are between $\frac{3}{5}$ and $\frac{4}{5}$

The fact that fractions are "dense" in this way is one of the important mathematical characteristics that distinguishes them from whole numbers.

Fraction Operations

Adding and Subtracting Fractions

As students add and subtract fractions, an appropriate sequence might be

- fractions less than 1 with the same denominator, where the sum, or the minuend, is less than 1, like $\frac{3}{8} + \frac{4}{8}$ or $\frac{5}{6} - \frac{2}{6}$
- any fractions less than 1 with the same denominator, like $\frac{5}{8} + \frac{7}{8}$
- any fractions less than 1 with a sum, or minuend, less than 1, like $\frac{1}{4} + \frac{2}{3}$ or $\frac{5}{6} - \frac{3}{8}$
- any fractions less than 1, like $\frac{3}{5} + \frac{7}{8}$
- any improper and mixed fractions, like $4\frac{3}{5} + 2\frac{7}{8}, \frac{5}{3} + \frac{8}{7}$, or $2\frac{2}{5} - \frac{9}{7}$

Whenever calculating sums and differences, students should also estimate using benchmarks by asking, for example, is the sum closer to 1 or $\frac{1}{2}$? Is the difference closer to 0 or $\frac{1}{2}$?

It is also important that students are exposed to different meanings of subtraction—sometimes take away, sometimes comparison, and sometimes missing addend—and to the relationship between addition and subtraction.

It makes sense to expose students to addition and subtraction of improper fractions after mixed numbers have been introduced, since then it is possible for students to check the reasonableness of the sum or difference by estimating with the whole number values.

One way to begin the topic of fraction addition and subtraction is to start with what students already know about representing a fraction. For example, students can begin with a representation of $\frac{5}{8}$ as 5 parts out of 8 and then they can colour the representation to model different addition expressions:

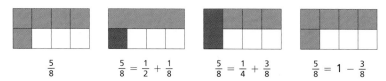

$\frac{5}{8}$ $\frac{5}{8} = \frac{1}{2} + \frac{1}{8}$ $\frac{5}{8} = \frac{1}{4} + \frac{3}{8}$ $\frac{5}{8} = 1 - \frac{3}{8}$

Once students see that $\frac{5}{8}$ can be represented as $\frac{1}{2}$ and $\frac{1}{8}$, it makes sense that $\frac{5}{8} = \frac{1}{2} + \frac{1}{8}$.

Different models and approaches can and should be used to show addition and subtraction of fractions. Some examples are shown on the next page.

MODEL 1 Think of the denominator as a unit when fractions have the same denominator.

ADDING	SUBTRACTING
For example, fifths are just like any other unit. Three of them plus one more of them is four of them altogether.	For example, eighths are just like any other unit. If you have five of them and you take away two of them, you have three of them left.
$\frac{3}{5} + \frac{1}{5}$ means 3 fifths + 1 fifth.	$\frac{5}{8} - \frac{2}{8}$ means 5 eighths − 2 eighths.
Since 3 fifths + 1 fifth = 4 fifths, then $\frac{3}{5} + \frac{1}{5} = \frac{4}{5}$.	Since 5 eighths − 2 eighths = 3 eighths, then $\frac{5}{8} - \frac{2}{8} = \frac{3}{8}$.

MODEL 2 Use number lines.

Adding The sum of $\frac{3}{5} + \frac{1}{5}$ is the resulting position when you start at $\frac{3}{5}$ on a number line and add $\frac{1}{5}$.

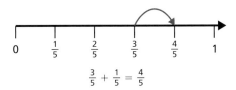

$$\frac{3}{5} + \frac{1}{5} = \frac{4}{5}$$

Subtracting The difference for $\frac{3}{5} - \frac{1}{5}$ can have these two meanings:

Missing addend subtraction The distance from $\frac{1}{5}$ to $\frac{3}{5}$ on a number line, represented below by the length of the arrow.

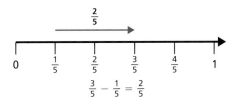

$$\frac{3}{5} - \frac{1}{5} = \frac{2}{5}$$

Take away subtraction The resulting position when you start at $\frac{3}{5}$ on a number line and move left $\frac{1}{5}$.

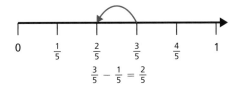

$$\frac{3}{5} - \frac{1}{5} = \frac{2}{5}$$

The context of the problem might suggest which approach to use, but because the meanings are equivalent, either approach is correct in any subtraction situation. For example:

- If the problem were: *I finished $\frac{1}{5}$ of my report. I wanted to have $\frac{3}{5}$ done by Friday. What fraction of the report do I have left to do by Friday?*, the first model might make more sense.
- If the problem were: *I had completed $\frac{3}{5}$ of my homework. I realized that I had done $\frac{1}{5}$ of it incorrectly. What fraction of my homework did I complete correctly?*, the second model fits better.

Teaching Idea | **2.37**

To reinforce BIF 5, begin with whole numbers and ask: *Why might you think of 303 − 299 as the distance from 299 to 303 but think of 303 − 4 as the number that is 4 less than 303?* [From 299 to 303 is not very far, so it is quick to count up, but the distance between 4 and 303 is great, so it is quicker to count back 4 from 303.] *Then ask: Which strategy would you use to subtract $6\frac{1}{3} - \frac{2}{3}$?* [e.g., count up from $\frac{2}{3}$ to 1 and then add $5\frac{1}{3}$]

MODEL 3 Use fraction strips.

To model addition and subtraction of fractions, you can use fractions strips.

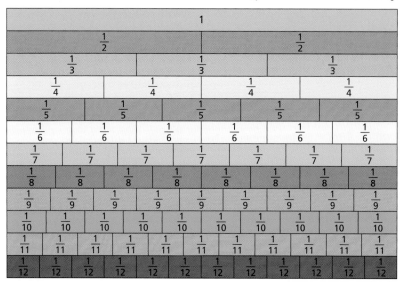

Fraction strips are useful for modelling fraction addition and subtraction because they are an ideal manipulative for showing equivalent fractions.

Adding The sum of $\frac{2}{3} + \frac{1}{4}$ is the length of a fraction strip that has the same total length as the $\frac{2}{3}$ strip and the $\frac{1}{4}$ strip placed end to end.

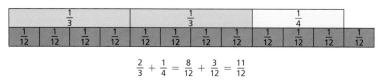

$$\frac{2}{3} + \frac{1}{4}$$

If you compare it to the 1, or whole strip, it you can see that the sum is less than 1, but it is difficult to see the exact sum.

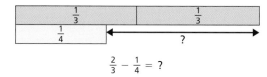

$$\frac{2}{3} + \frac{1}{4} < 1$$

It is important for students to be comfortable with equivalent fractions and common multiples before adding and subtracting fractions symbolically.

To determine the exact sum, students can use the idea that thirds and fourths can both be represented as twelfths (twelfths because 12 is a common multiple of the two denominators, 3 and 4). This is modelled using fraction strips for twelfths.

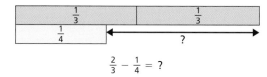

$$\frac{2}{3} + \frac{1}{4} = \frac{8}{12} + \frac{3}{12} = \frac{11}{12}$$

Subtracting The difference for $\frac{2}{3} - \frac{1}{4}$ can be interpreted using the comparison meaning of subtraction, that is, how much longer a $\frac{2}{3}$ fraction strip is than a $\frac{1}{4}$ fraction strip.

$$\frac{2}{3} - \frac{1}{4} = ?$$

You can see that the difference is a bit more than $\frac{1}{3}$.

Teaching Idea | **2.38**

To reinforce BIF 3, that renaming fractions can help simplify calculations, ask: *Which of these computations would be easy to do without using equivalent fractions? Which would not? Explain.*

$\frac{3}{8} + \frac{2}{8}$ $\frac{4}{9} - \frac{2}{9}$ $\frac{5}{6} + \frac{3}{4}$

[The first two would be easy to do since you can just add 3 eighths + 2 eighths or take away 2 ninths from 4 ninths mentally, but it would be hard to do $\frac{5}{6} + \frac{3}{4}$ without equivalent fractions.]

The exact difference can be found using the fraction strips for twelfths.

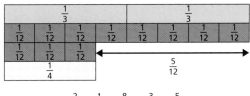

$$\frac{2}{3} - \frac{1}{4} = \frac{8}{12} - \frac{3}{12} = \frac{5}{12}$$

MODEL 4 Use grids and counters.

To model addition and subtraction of fractions, you can use a grid and counters.

Adding $\frac{3}{5} + \frac{1}{3}$

Step 1 To model $\frac{3}{5} + \frac{1}{3}$, begin with a 3 by 5 grid.

You can use each row to model one third and each column to model one fifth.

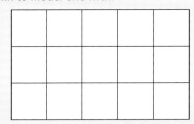

A 3 row by 5 column grid

Step 2 Model $\frac{3}{5}$ by filling 3 of the 5 columns with counters.

The grid and counters show that $\frac{3}{5} = \frac{9}{15}$.

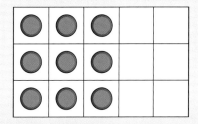

$\frac{3}{5}$ or $\frac{9}{15}$ of the grid has blue counters.

Step 3 To add $\frac{1}{3}$, you want to fill 1 of the 3 rows with a different colour counter.

First move the blue counters in one of the rows to an empty part of the grid.

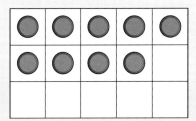

Fill the empty row with green counters.
The grid and counters show that $\frac{1}{3} = \frac{5}{15}$.

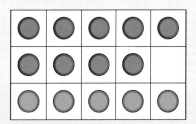

$\frac{14}{15}$ of the grid is full.

$$\frac{3}{5} + \frac{1}{3} = \frac{9}{15} + \frac{5}{15} = \frac{14}{15}$$

Step 1 To model $\frac{3}{4} - \frac{2}{3}$ begin with a 3 by 4 grid.

You can use each row to model one third and each column to model one fourth.

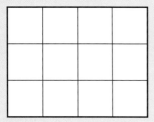

A 3 row by 4 column grid

Step 2 Model $\frac{3}{4}$ by filling 3 of the 4 columns with counters.

The grid and counters show that $\frac{3}{4} = \frac{9}{12}$.

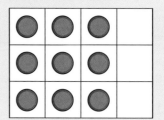

Step 3 To subtract $\frac{2}{3}$, you want to take away all the counters in 2 rows (since 2 rows is $\frac{2}{3}$ of the grid).

First move the blue counters to fill 2 rows of the grid.

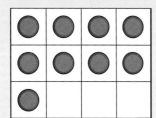

Remove the counters from 2 rows.

The grid and counters show that $\frac{2}{3} = \frac{8}{12}$.

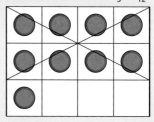

$\frac{1}{12}$ of the grid is left with a counter.

$$\frac{3}{4} - \frac{2}{3} = \frac{9}{12} - \frac{8}{12} = \frac{1}{12}$$

Adding and Subtracting Mixed Numbers

When adding and subtracting mixed numbers, the same models can be used as with proper fractions. Normally, the whole-number amounts are added or subtracted separately from the fraction amounts, but there are exceptions.

Adding and Subtracting Whole Numbers and Fractions

Adding a whole number and a proper fraction is simply combining them to form a mixed number. For example, $5 + \frac{2}{3} = 5\frac{2}{3}$. Subtracting a fraction from a whole number can be a bit more complicated and strategies are needed for this.

Students might visualize a number line when mentally subtracting a fraction from a whole number.

Subtracting $1 - \frac{2}{3}$

USING A NUMBER LINE	USING A GRID AND COUNTERS
The difference is the distance from $\frac{2}{3}$ to 1.	The difference is how much of the grid is filled after $\frac{2}{3}$ is taken away

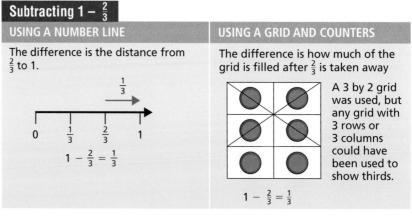

Subtracting $4 - \frac{2}{3}$

USING A NUMBER LINE	USING EQUIVALENT FRACTIONS
The difference is the combined distance from $\frac{2}{3}$ to 1 and from 1 to 4. 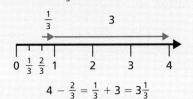 $4 - \frac{2}{3} = \frac{1}{3} + 3 = 3\frac{1}{3}$	$4 - \frac{2}{3} = \frac{12}{3} - \frac{2}{3}$ $= 12 \text{ thirds} - 2 \text{ thirds}$ $= 10 \text{ thirds}$ $= \frac{10}{3}$ $= 3\frac{1}{3}$

Teaching Idea | **2.39**

To connect fraction and whole-number calculations (BIF 5), ask students how they might subtract 400 − 193 mentally. Once someone suggests counting up from 193 to 200 (7) and then from 200 to 400 (200), which is 207 altogether (400 − 193 = 207), ask: *How could you use that idea to calculate $3 - 1\frac{5}{6}$? Sketch a number line to show what it might look like.*

Adding and Subtracting Mixed Numbers

Adding/Subtracting Whole Numbers and Fractions Separately

$3\frac{4}{5} + 1\frac{1}{2}$	$3\frac{4}{5} - 1\frac{1}{2}$
$3\frac{4}{5} + 1\frac{1}{2} = 3 + 1 + \frac{4}{5} + \frac{1}{2}$ $= 4 + \frac{4}{5} + \frac{1}{2}$ $= 4 + \frac{8}{10} + \frac{5}{10}$ $= 4 + \frac{13}{10}$ $= 4 + 1\frac{3}{10}$ $= 5\frac{3}{10}$	$3\frac{4}{5} - 1\frac{1}{2} = (3 - 1) + (\frac{4}{5} - \frac{1}{2})$ $= 2 + (\frac{4}{5} - \frac{1}{2})$ $= 2 + (\frac{8}{10} - \frac{5}{10})$ $= 2 + \frac{3}{10}$ $= 2\frac{3}{10}$ This approach works in this situation because $\frac{4}{5} > \frac{1}{2}$.

Note that subtracting whole numbers and fraction parts separately and then adding the differences only works when the first fraction (in the minuend) is greater than the second fraction (in the subtrahend).

Subtracting $3\frac{1}{4} - 1\frac{3}{5}$

USING A NUMBER LINE	RENAMING AND SUBTRACTING
The difference is the combined distance from $1\frac{3}{5}$ to $3\frac{1}{4}$. 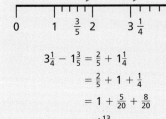 $3\frac{1}{4} - 1\frac{3}{5} = \frac{2}{5} + 1\frac{1}{4}$ $= \frac{2}{5} + 1 + \frac{1}{4}$ $= 1 + \frac{5}{20} + \frac{8}{20}$ $= 1\frac{13}{20}$	Since $\frac{1}{4} < \frac{3}{5}$, you cannot subtract the whole number and fractions parts separately and then add the differences. But, if you rename $3\frac{1}{4}$ so it has a fraction greater than $\frac{3}{5}$, you can. $3\frac{1}{4} = 2 + \frac{4}{4} + \frac{1}{4} = 2\frac{5}{4}$ $2\frac{5}{4} - 1\frac{3}{5} = (2 - 1) + (\frac{5}{4} - \frac{3}{5})$ $= 1 + \frac{25}{20} - \frac{12}{20}$ $= 1\frac{13}{20}$

Multiplying and Dividing Fractions

As students multiply and divide fractions, an appropriate sequence of types of problems might be

- multiplying a fraction by a whole number, e.g., $5 \times \frac{2}{3}$
- multiplying two fractions less than 1, e.g., $\frac{4}{5} \times \frac{2}{3}$
- dividing fractions by whole numbers, e.g., $\frac{4}{5} \div 6$
- dividing a fraction less than 1 by a lesser one, e.g., $\frac{3}{4} \div \frac{1}{3}$
- dividing any two fractions, e.g., $\frac{1}{8} \div \frac{2}{5}$
- dividing mixed numbers, e.g., $4\frac{2}{3} \div 1\frac{3}{4}$

For students, one of the difficult aspects of multiplying two fractions is the fact that the product can be less than either of the factors; this is something that they have never experienced before with whole numbers. So it is very important to remind students of what multiplication really means. You can do this using different multiplication meanings and models.

Multiplying a Fraction by a Whole Number

When students begin multiplication with fractions, they can use the repeated addition meaning of multiplication to interpret the multiplication of a fraction by a whole number. For example, $4 \times \frac{3}{5}$ means 4 sets of $\frac{3}{5}$, so $4 \times \frac{3}{5} = \frac{3}{5} + \frac{3}{5} + \frac{3}{5} + \frac{3}{5} = \frac{12}{5} = 2\frac{2}{5}$.

Multiplying Two Fractions

Multiplication as Equal Sets Using Fraction Strips

One of the meanings of multiplication is the equal groups, or equal sets, meaning (using the word "of"). It helps to use the meaning to connect whole-number multiplication to fraction multiplication. For example, 3×4 means 3 sets of 4. It only makes sense that $\frac{1}{2} \times \frac{3}{5}$ means $\frac{1}{2}$ set of $\frac{3}{5}$. Using fraction strips, students can see that $\frac{1}{2} \times \frac{3}{5}$ is $\frac{1}{2}$ of $\frac{3}{5}$, or $\frac{3}{10}$.

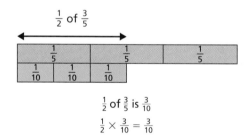

$\frac{1}{2}$ of $\frac{3}{5}$ is $\frac{3}{10}$

$\frac{1}{2} \times \frac{3}{10} = \frac{3}{10}$

Multiplication as Area Using Grids

Another meaning of multiplication that students used with whole numbers is the notion that multiplication represents the area of a rectangle with given dimensions. That concept can be used to model, for example, $\frac{3}{5} \times \frac{2}{3}$, as shown at the top of the next page.

A rectangle is created with one dimension $\frac{3}{5}$ and the other $\frac{2}{3}$. This can be done using a 5 by 3 grid (since the denominators are 5 and 3), or by folding a square piece of paper to form thirds in one direction and fifths in the other.

A common misconception about multiplication is that it always makes things bigger. Students often have a similar misconception about division—that it always makes things smaller. These misconceptions can be addressed by focusing on the meanings of the operations and using models.

Teaching Idea | **2.40**

Before showing students that $3 \times \frac{2}{3} = \frac{2}{3} + \frac{2}{3} + \frac{2}{3}$, introduce a problem like this:

Rebecca exercised $\frac{2}{3}$ h each day for 3 days. How long did she exercise?

Students can solve it using minutes (whole numbers) [3 × 40 min = 120 min] or hours [$3 \times \frac{2}{3}$ h = 2 h].

After students share their solutions, focus on BIF 5 by asking: *Why could you have solved it using $\frac{2}{3} + \frac{2}{3} + \frac{2}{3}$? How is that like multiplying whole numbers?* [3 × 2 = 2 + 2 + 2, so it makes sense that $3 \times \frac{2}{3} = \frac{2}{3} + \frac{2}{3} + \frac{2}{3}$] *Why was only the numerator multiplied by 3?* [You can think of $\frac{2}{3}$ as 2 thirds, so you have 3 × 2 thirds, which is 6 thirds.]

USING A GRID MODEL TO MULTIPLY

Create a grid with dimensions equal to the denominators.	Model each factor as the dimension of a rectangle.	Create a rectangle with those dimensions.

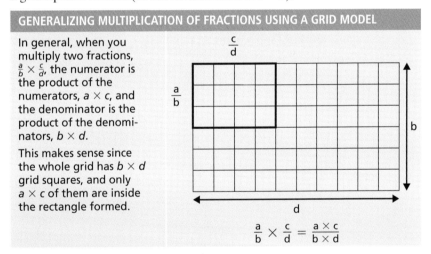

On a 3 by 5 grid, fifths can be modelled along one dimension and thirds along the other dimension.

The dimensions of a rectangle that is $\frac{3}{5}$ by $\frac{2}{3}$.

$$\frac{3}{5} \times \frac{2}{3} = \frac{6}{15}$$

The area of the blue rectangle is 6 out of 15 grid squares, or $\frac{6}{15}$.

After working with a model such as the grid model, students will observe that the numerator of the product is the product of the numerators of the factors and the denominator of the product is the product of the denominators of the factors. For example, in the grid model above, $\frac{3}{5} \times \frac{2}{3} = \frac{2 \times 3}{5 \times 3} = \frac{6}{15}$. This makes sense, since the blue rectangle has 3 columns with 2 grid squares in each (so the numerator is 3×2), and the whole grid is made up of 5 columns with 3 grid squares in each (so the denominator is 5×3).

GENERALIZING MULTIPLICATION OF FRACTIONS USING A GRID MODEL

In general, when you multiply two fractions, $\frac{a}{b} \times \frac{c}{d}$, the numerator is the product of the numerators, $a \times c$, and the denominator is the product of the denominators, $b \times d$.

This makes sense since the whole grid has $b \times d$ grid squares, and only $a \times c$ of them are inside the rectangle formed.

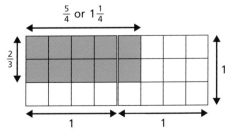

$$\frac{a}{b} \times \frac{c}{d} = \frac{a \times c}{b \times d}$$

This grid model can also be used to multiply fractions greater than 1. For example, to show $\frac{2}{3} \times \frac{5}{4}$, you would use two 3 by 4 grids and then create a rectangle with dimensions $\frac{5}{4}$ by $\frac{2}{3}$ using both grids.

The green rectangle covers 10 grid squares.
Since 12 grid squares make a whole, $\frac{2}{3} \times \frac{5}{4} = \frac{10}{12}$.

Teaching Idea | 2.41

Before showing students a grid model for multiplying fractions, ask them to explain how they would calculate the area of a rectangle. [The area is the product of the two dimensions.] Then reverse the process and ask them why 3×4 could be modelled as the area of a 3 by 4 rectangle. [3×4 is 3 rows of 4 and that is what you would get if you covered a 3 by 4 rectangle with unit square tiles.]

To connect whole-number and fraction operations (BIF 5), ask: *What rectangle would you draw to show $\frac{2}{3} \times \frac{4}{5}$?* [a rectangle that is $\frac{2}{3}$ of a unit by $\frac{3}{5}$ of a unit] *How could you use your picture to figure out the area of your rectangle?* [See what fraction of a 1 by 1 rectangle the smaller rectangle is.]

Teaching Idea | 2.42

Ask students why it would be easy to use fraction strips to show $\frac{2}{3} \times \frac{3}{4}$, but not $\frac{2}{3} \times \frac{4}{5}$.

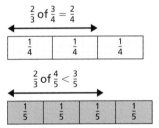

$\frac{2}{3}$ of $\frac{3}{4} = \frac{2}{4}$

$\frac{2}{3}$ of $\frac{4}{5} < \frac{3}{5}$

Show that different models and strategies are more useful in different situations (BIF 4) by asking: *How is this like using coloured counters to subtract $(-4) - (-2)$ as compared to $(-4) - (+3)$?* [You can take away 2 blue counters from 4 blue counters but you can't take away 3 red counters from 4 blue ones.]

Revisiting Equivalent Fractions

Once students understand how to multiply fractions, it is possible to revisit the concept of creating equivalent fractions in a new way. Students know that to create an equivalent fraction, they can multiply the numerator and denominator by the same amount, for example, $\frac{2}{3} = \frac{6}{9}$ because $\frac{2 \times 3}{3 \times 3} = \frac{6}{9}$. Now they can see that they have also multiplied two fractions $\frac{2}{3} \times \frac{3}{3} = \frac{6}{9}$. And, since $\frac{3}{3} = 1$, they have actually multiplied $\frac{2}{3}$ by 1, which they know does not change the value.

Multiplying Mixed Numbers

Students can multiply mixed numbers by using an area model or by symbolically multiplying the equivalent improper fractions and applying the algorithm $\frac{a}{b} \times \frac{c}{d} = \frac{a \times c}{b \times d}$.

Multiplying $2\frac{1}{2} \times 3\frac{1}{3}$

USING AN AREA MODEL	SYMBOLICALLY USING IMPROPER FRACTIONS
$2\frac{1}{2} \times 3\frac{1}{3}$ is the area of a rectangle with dimensions $2\frac{1}{2}$ by $3\frac{1}{3}$.	Change each mixed number to an improper fraction and then apply the algorithm $\frac{a}{b} \times \frac{c}{d} = \frac{a \times c}{b \times d}$.

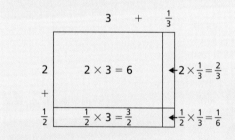

Area $= 2\frac{1}{2} \times 3\frac{1}{3} = 6 + \frac{2}{3} + \frac{3}{2} + \frac{1}{6}$ $= 6 + \frac{4}{6} + \frac{9}{6} + \frac{1}{6}$ $= 6 + \frac{14}{6}$ $= 6 + 2\frac{2}{6}$ $= 8\frac{2}{6}$	$2\frac{1}{2} \times 3\frac{1}{3} = \frac{5}{2} \times \frac{10}{3}$ $= \frac{50}{6}$ $= 8\frac{2}{6}$ If students use this approach, they should be encouraged to use mixed-number equivalents to estimate the answer to see if their product is reasonable. $2\frac{2}{3} \times 3\frac{1}{2}$ is about $3 \times 3 = 9$, so a product of $8\frac{2}{6}$ seems reasonable.

Dividing Fractions

Dividing a Fraction by a Whole Number

Thinking about division as sharing can help students understand division of fractions by whole numbers. For example, $\frac{6}{7} \div 3$ can means 6 sevenths is shared by 3 people. Each person gets 2 sevenths, so $\frac{6}{7} \div 3 = \frac{2}{7}$. Students will observe that the numerator is divided by the whole number and the denominator does not change ($\frac{6}{7} \div 3 = \frac{6 \div 3}{7} = \frac{2}{7}$).

Sometimes, the division is not as straightforward. For example, even though $\frac{6}{7} \div 4$ can be thought of as 6 sevenths shared by 4 people so that each person gets $1\frac{1}{2}$ sevenths, this is awkward to write. If an equivalent fraction for $\frac{6}{7}$ is created with a numerator that is divisible by 4 (the divisor), the division is much easier to grasp and the resulting quotient is a simple fraction.

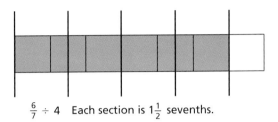

$\frac{6}{7} \div 4$ Each section is $1\frac{1}{2}$ sevenths.

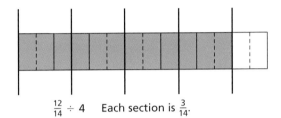

$\frac{12}{14} \div 4$ Each section is $\frac{3}{14}$.

Dividing Two Fractions

Some division of fractions can be easy for students without knowing any formal procedures. This is particularly the case if the division is "read" to them meaningfully. For example, reading $\frac{1}{2} \div \frac{1}{4}$ as "How many 1 fourths are in $\frac{1}{2}$?" allows students who do not know any formal procedure to calculate a quotient of 2; all they need to know is that $\frac{1}{2} = \frac{2}{4}$ (one half is two fourths).

With the use of manipulatives and a recognition that $a \div b$ means how many bs are in a, students can easily estimate quotients. Just looking at the fraction strips, it is clear that $\frac{2}{3} \div \frac{1}{10}$ is a bit less than 7 because there are almost 7 tenths in $\frac{2}{3}$.

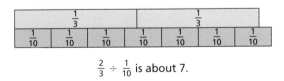

$\frac{2}{3} \div \frac{1}{10}$ is about 7.

To get the exact answer, though, students need to use more exact procedures, such as those described below.

Algorithm 1 Using Common Denominators Among the easiest types of fraction division for students are questions like $\frac{6}{7} \div \frac{2}{7}$ (How many 2 sevenths are in 6 sevenths?) or $\frac{8}{9} \div \frac{2}{9}$ ("How many 2 ninths are in 8 ninths?"), because the denominators are the same and one numerator is a factor of the other. Students readily see that the quotient is simply the quotient of the numerators; the denominator is not used in the computation ($\frac{6}{7} \div \frac{2}{7} = 6 \div 2 = 3$).

The question is only slightly more difficult if the denominators are the same but one numerator is not a factor of another, for example, $\frac{3}{5} \div \frac{2}{5}$.

Teaching Idea | **2.43**

To reinforce that using an alternative representation can make a calculation simpler (BIF 3), ask: *Why might it be easier to calculate 297 ÷ 3 mentally by thinking of 297 as 300 − 3?* [It changes it into two easy divisions plus a simple subtraction: 300 ÷ 3 = 100, 3 ÷ 3 = 1, and 100 − 1 = 99] *Why might it be easier to calculate $\frac{12}{15} \div 3$ instead of $\frac{4}{5} \div 3$ using mental math?* [The first quotient is $\frac{12 \div 3}{15}$, but it's not obvious what the second quotient is since 4 is not divisible by 3.] *What about $\frac{10}{15} \div \frac{2}{5}$ instead of $\frac{2}{3} \div \frac{2}{5}$?* [It's easier because the first quotient is $\frac{10 \div 2}{15 \div 5}$, but 3 is not divisible by 5, so the second quotient is not as obvious.]

The calculation $\frac{3}{5} \div \frac{2}{5}$ is asking "How many 2s (in this case, 2 fifths) are in 3 (in this case, 3 fifths)?"

The diagram to the right shows that 1 set of 2 fifths is not enough and 2 sets of 2 fifths is too much, but $1\frac{1}{2}$ sets of 2 fifths is perfect, so $\frac{3}{5} \div \frac{2}{5} = 1\frac{1}{2}$.

Notice that $1\frac{1}{2}$ is the quotient of the numerators ($3 \div 2 = 1\frac{1}{2}$).

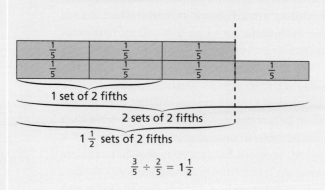

1 set of 2 fifths

2 sets of 2 fifths

$1\frac{1}{2}$ sets of 2 fifths

$$\frac{3}{5} \div \frac{2}{5} = 1\frac{1}{2}$$

Dividing using common denominators can be generalized as shown below:

$$\frac{a}{c} \div \frac{b}{c} = a \div b$$

This works even when the fractions have different denominators, since you can always create equivalent fractions with a common denominator; e.g.,

$$\frac{3}{4} \div \frac{2}{3} = \frac{9}{12} \div \frac{8}{12}$$
$$= 9 \div 8$$
$$= \frac{9}{8} \text{ or } 1\frac{1}{8}$$

Teaching Idea | 2.44

We introduce the reciprocal to students for fraction division, but the concept has very little meaning to them.

One way to give it meaning as well as bring out BIF 2 is to ask questions such as: *If the red trapezoid pattern block represents 1, what does the blue rhombus represent?* [$\frac{2}{3}$] *If the blue rhombus represents 1, what does the red trapezoid represent? What do you notice?* [$\frac{3}{2}$; the two answers are related, the numerator and denominator are just switched]

Repeat with other blocks such as the yellow hexagon and red trapezoid [$\frac{1}{2}$ and 2 or $\frac{2}{1}$], and 2 blue rhombuses and 3 green triangles [$\frac{3}{4}$ and $\frac{4}{3}$].

Algorithm 2 Inverting and Multiplying The more traditional algorithm for dividing fractions is inverting the divisor (using its reciprocal) and then multiplying. For example, $\frac{3}{4} \div \frac{2}{3} = \frac{3}{4} \times \frac{3}{2} = \frac{9}{8}$. This algorithm can be explained in several ways. One meaningful approach is shown below:

$1 \div \frac{1}{3} = 3$, since there are 3 thirds in 1.

$\frac{3}{4} \div \frac{1}{3} = \frac{3}{4}$ of 3, since there can only be $\frac{3}{4}$ as many thirds in $\frac{3}{4}$ as there are in 1.

$\frac{3}{4}$ of $3 = \frac{3}{4} \times 3$

$\frac{3}{4} \div \frac{2}{3} = \frac{1}{2}$ of ($\frac{3}{4} \times 3$) since $\frac{2}{3}$ is twice as much as $\frac{1}{3}$, so only half as many $\frac{2}{3}$s will fit into a whole as $\frac{1}{3}$s.

$\frac{3}{4} \div \frac{2}{3} = \frac{1}{2} \times (\frac{3}{4} \times 3) = \frac{3}{4} \times \frac{1}{2} \times 3 = \frac{3}{4} \times \frac{3}{2}$

So, $\frac{3}{4} \div \frac{2}{3} = \frac{3}{4} \times \frac{3}{2}$.

The invert and multiply is the better-known algorithm, probably because it is easier to use in algebraic situations in later grades, but either the common denominator or the invert and multiply algorithms are appropriate for students to use.

Dividing Mixed Numbers

The easiest way to divide mixed numbers is to use the equivalent improper fractions and apply one of the division algorithms to these values, as shown below. When students work symbolically like this, they should estimate to check if their answer makes sense.

Dividing $2\frac{1}{2} \div 1\frac{1}{3}$		
INVERT AND MULTIPLY	**COMMON DENOMINATOR**	**ESTIMATING TO CHECK**
$2\frac{1}{2} \div 1\frac{1}{3}$	$2\frac{1}{2} \div 1\frac{1}{3}$	$2\frac{1}{2} \div 1\frac{1}{3}$
$= \frac{5}{2} \div \frac{4}{3}$	$= \frac{5}{2} \div \frac{4}{3}$	$\approx 2\frac{1}{2} \div 1\frac{1}{4}$
$= \frac{5}{2} \times \frac{3}{4}$	$= \frac{15}{6} \div \frac{8}{6}$	≈ 2
$= \frac{15}{8}$	$= 15 \div 8$	Since $1\frac{1}{4}$ is a bit smaller than $1\frac{1}{3}$, the estimate is a bit high.
$= 1\frac{7}{8}$	$= \frac{15}{8}$	
	$= 1\frac{7}{8}$	

Decimals

Decimals are not unfamiliar to students who have seen prices in dollar and cents from even before they get to school. However, students' understanding of this notation, for example, $3.14, is in terms of 3 dollars and 14 cents and not 3 dollars and 14 hundredths of a dollar, which is the underlying decimal meaning.

BIG IDEAS FOR DECIMALS

1. Decimals are an alternative representation to fractions, but one that allows for modelling, comparisons, and calculations that are consistent with whole numbers, because decimals extend the pattern of the base ten place value system.

2. A decimal can be read and interpreted in different ways; sometimes one representation is more useful than another in interpreting or comparing decimals or for performing and explaining a computation.

Each teaching idea in this section of the chapter will indicate which Big Idea(s) for Decimals (BID) can be emphasized.

Representing Decimals

It is not really possible for students to make sense of decimals without some understanding of fractions. An introduction to decimals requires familiarity with the concept of fraction tenths. It is natural to begin with items that come in tens to begin renaming fraction tenths as decimal tenths.

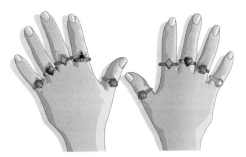

$\frac{2}{10}$, or 0.2, of the fingers have blue rings.
Fingers are familiar and they come in tens, so they make an excellent introductory context for decimals.

Some students who are fairly comfortable with decimal tenths and hundredths become less comfortable with decimal thousandths. It is important not to hurry the introduction of thousandths until students are ready, and always use concrete or pictorial support to represent thousandths.

A good pictorial representation for tenths, hundredths, and thousandths is a square model, as shown at the top of the next page.

In this next example, the base ten block that is used to represent 100 when modelling whole numbers (the flat) is now used to represent the whole, or 1. For some students, this causes confusion, and using a new colour of block (if available) or a different model altogether might be helpful.

Representing 2.35 with Base Ten Blocks Different Ways

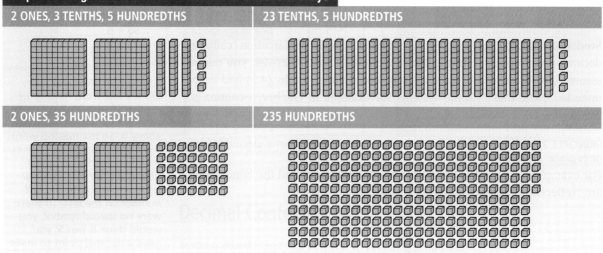

| 2 ONES, 3 TENTHS, 5 HUNDREDTHS | 23 TENTHS, 5 HUNDREDTHS |
| 2 ONES, 35 HUNDREDTHS | 235 HUNDREDTHS |

Teaching Idea | 2.47

Have students fill a 10 by 10 grid with pennies to show that dimes are worth 0.10 or 0.1 or $\frac{1}{10}$ of a dollar (one row is worth a dime), and pennies are 0.01 or $\frac{1}{100}$ of a dollar (one grid square is worth a penny). They will also see that 0.20 is equivalent to 0.2 or $\frac{2}{10}$, since 20 pennies (0.20) fill two columns.

Show the usefulness of equivalent representations (BID 2) by asking: *Why is it easier to compare $\frac{1}{2}$ and $\frac{1}{5}$ when they are renamed as $\frac{5}{10}$ and $\frac{2}{10}$?* [You just compare the numerators.] *When might it be easier to think of 0.2 as 0.20, or 20 hundredths, to compare it to another decimal?* [e.g., if I was comparing 0.2 to 0.34, I would think 20 hundredths is less than 34 hundredths.]

PRINCIPLE 5 Decimals can be renamed as other decimals or fractions.

Just like fractions, decimals have multiple names, including both fraction and decimal equivalents.

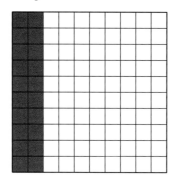

20 out of 100 squares of this grid are red.

$\frac{20}{100}$, 0.20, $\frac{2}{10}$, 0.2, or $\frac{1}{5}$

Equivalent Decimals

Explaining Why 0.2 = 0.20

USING FRACTION EQUIVALENTS	USING PLACE VALUE
One way to explain equivalence of decimals is to use fraction equivalents for each decimal. $0.2 = \frac{2}{10}$ $\frac{2}{10} = \frac{2}{10} \times \frac{10}{10}$ $\quad = \frac{20}{100}$ $\frac{20}{100} = 0.20$ Therefore 0.2 = 0.20. This approach only works once students have learned how to multiply fractions.	Another way to explain equivalence is to use place value language and expanded notation. 0.2 = 2 tenths 0.20 = 2 tenths + 0 hundredths Since 0 hundredths has no effect, 0.20 must be the same as 0.2. Therefore 0.2 = 0.20. This approach is useful as soon as students learn what tenths and hundredths are.

Equivalent Decimals and Precision

It should be noted that when you speak about measurement units, you cannot equate, for example, 3.2 m and 3.20 m, since different levels of precision are implied by the use of more decimal places. The measurement 3.20 m indicates that the length could be anywhere between 3.195 m and 3.205 m (the range of values that could be rounded to 3.20), whereas 3.2 m could be anywhere between 3.15 m and 3.25 m (the range of values that could be rounded to 3.2), which makes 3.2 m less precise than 3.20 m.

Equivalent Fractions and Decimals

Relating Decimals to Familiar Unit Fractions

Decimals are introduced as tenths or hundredths, so students immediately recognize the relationship between decimals and fraction tenths or hundredths; for example, $\frac{2}{10}$ is 0.2 and 0.34 is $\frac{34}{100}$. But often it is convenient to take advantage of other fraction and decimal relationships. For example,

- 0.5 or 0.50 is another name for $\frac{1}{2}$
- 0.25 is another name for $\frac{1}{4}$
- 0.125 is another name for $\frac{1}{8}$
- 0.333... is another name for $\frac{1}{3}$

Knowing common fraction–decimal relationships can help students interpret decimals meaningfully. For example, they see 0.48 and realize that it is almost $\frac{1}{2}$.

Explaining Equivalent Fractions and Decimals

USING A MONEY MODEL	USING DIVISION

USING A MONEY MODEL

Since a quarter is 25¢, or $\frac{25}{100}$ of a dollar, and 4 quarters make a dollar, then 0.25 = $\frac{1}{4}$.

Since a dime is 10¢, or $\frac{10}{100}$ of a dollar, and 10 dimes make a dollar, then 0.10 = $\frac{1}{10}$.

USING DIVISION

You can think of $\frac{1}{2}$ as 1 ÷ 2. When you divide 1 by 2 on a calculator, the display reads 0.5. You read this as "5 tenths" and realize that this makes sense since 5 out of 10 is half.

$$1 \;\blacksquare\; 2 \;\blacksquare\; \boxed{\qquad 0.5}$$

0.5 = $\frac{1}{2}$ makes sense because 1 ÷ 2 = 0.5 or 5 tenths, and 5 tenths out of 10 tenths is $\frac{1}{2}$.

More Complex Fraction–Decimal Relationships

In about Grade 7 or 8, students are asked to apply the fraction–decimal relationship even when fractions are not as easy to interpret as decimals. For example, students are asked to think of $\frac{1}{3}$ as a decimal. In general, you use the division meaning of fraction to make this make sense. Since $\frac{1}{3}$ = 1 ÷ 3, students can use either a calculator or a pencil-and-paper algorithm to see that $\frac{1}{3}$ = 0.3333....

Students will observe that, unlike other decimals they have met before, this decimal repeats and does not terminate. Students will accept this because the calculator shows it, but they can use logical reasoning to make sense of it:

- Imagine dividing 1 whole so that 3 people can share it.
- Each person cannot get a whole, so rename the whole as 10 tenths.
- Each person gets 3 tenths, but 1 tenth is left over and cannot be shared.
- Rename the leftover 1 tenth as 10 hundredths.
- Each person gets 3 hundredths, but there is 1 hundredth left over that cannot be shared.
- Rename the leftover 1 hundredth as 10 thousandths, and so on.

Students see that every time there is 1 piece left over so the process will never end; every time, each person sharing gets 3 of the new and smaller unit, so $\frac{1}{3} = 0.333\ldots$ makes sense.

At this point, you usually introduce notation for a repeating decimal. For example, you can write $0.3333\ldots$ as $0.\overline{3}$ or as $0.\dot{3}$. In later grades, students learn how to convert a repeating decimal to a fraction.

Rounding and Estimating Decimals

Students should be able to round decimals to simpler decimals, such as 2.567 to 2.6 or 2.567 to 3. The conventions, or rules, for rounding are just like the ones for whole numbers. For example,

Rounding Conventions

DECIMAL	NEAREST THOUSANDTH	NEAREST HUNDREDTH	NEAREST TENTH	NEAREST WHOLE
2.9375	2.938	2.94	2.9	3
6.0693	6.069	6.07	6.1	6

Students should also estimate decimals using simple fractions; for example, 3.24 is almost $3\frac{1}{4}$, and 2.22 is a bit more than $2\frac{1}{5}$. To do this type of estimating, the common fraction–decimal equivalents below come in handy. Students can "round" the decimal to the nearest common fraction–decimal equivalent; for example, 5.81 is about 5.75, or $5\frac{3}{4}$.

Teaching Idea | **2.48**

To focus on BID 1, students can explore the decimal equivalents for these fractions:

$\frac{1}{9}, \frac{2}{9}, \frac{3}{9}, \ldots, \frac{8}{9}$

$[0.\overline{1}, 0.\overline{2}, 0.\overline{3}, \ldots, 0.\overline{9}]$

$\frac{1}{11}, \frac{2}{11}, \frac{3}{11}, \ldots, \frac{10}{11}$

$[0.\overline{09}, 0.\overline{18}, 0.\overline{27}, \ldots, 0.\overline{90}]$

$\frac{1}{7}, \frac{2}{7}, \frac{3}{7}, \ldots, \frac{6}{7}$

$[0.\overline{142857}, 0.\overline{285714},$ $0.\overline{428571}, \ldots, 0.\overline{857142}]$

Ask: *How does the fraction pattern for ninths show the values are increasing?* [Each time there is 1 more ninth.] *How does the decimal pattern show the increase is more than $\frac{1}{10}$?* [$\frac{1}{10}$ is 0.1; if you add 0.1 to 0.1111..., you get 0.2111..., but 0.2222... is more] *How does the pattern of increases show that $\frac{1}{9} > \frac{1}{10}$?* [Since $\frac{1}{9}$ was added, and you know more than $\frac{1}{10}$ was added, then $\frac{1}{9} > \frac{1}{10}$.]

Useful Fraction–Decimal Equivalents

$\frac{1}{8}$	$\frac{1}{5}$	$\frac{1}{4}$	$\frac{2}{5}$	$\frac{1}{2}$	$\frac{3}{5}$	$\frac{3}{4}$	$\frac{4}{5}$	$\frac{7}{8}$
0.125	0.2	0.25	0.4	0.5	0.6	0.75	0.8	0.875

Reading and Writing Decimals

Reading Decimals

Students should read a decimal like 3.2 as "3 and 2 tenths," not as "3 point 2" or "3 decimal 2." Reading a decimal like 7.23 as "7 and 23 hundredths" reveals the important connection between fractions and decimals, which the language "seven point two three" does not. However, people rarely read decimals with more than 3 digits after the decimal place in this meaningful way, although they could. For example, a decimal number like 0.4578 could be read as "four thousand five hundred seventy-eight ten thousandths," but is more likely to be read as "point four five seven eight." Once students are working comfortably with decimals of this many digits, the value of reading the decimal as a fraction is no longer as great.

Note that the decimal point is represented by the word "and." That is why it is recommended that students be taught to read whole numbers without using "and"; for example, it is preferable to read 547 as "five hundred forty-seven" rather than "five hundred and forty-seven," although using "and" in this case is not incorrect. As well, teachers and students should be careful to read the digit 0 as "zero," not "oh," to emphasize its mathematical value.

Writing Decimals

Decimals can be greater than 1 or less than 1. For this reason, encourage students to record decimals less than 1 using a zero in the ones place, for example, 0.2, rather than just .2. This reinforces that the decimal is less than 1 and eliminates the confusion when decimal points are written indistinctly. Note that, for assessment purposes, writing 0.2 as .2 is not incorrect unless the purpose is to determine if the student can apply the convention of recording a zero in the ones place. Numbers such as 3 can be written as 3 or as 3.0, depending on the context and level of precision required.

Comparing Decimals

Strategies for comparing decimals relate more closely to strategies for comparing whole numbers than to strategies for comparing fractions. This is, in fact, one of the reasons that a decimal system is used—to make comparison easy. The trick, as with whole numbers, is to ensure that values in the same places are being compared. In the examples below, each decimal pair is compared two different ways.

COMPARING DECIMALS USING PLACE VALUE DIFFERENT WAYS	
0.78 > 0.39, since 7 tenths > 3 tenths	43.8 > 8.27, since 4 tens > 0 tens
0.78 > 0.39, since 78 hundredths > 39 hundredths	43.8 > 8.27, since 43 ones > 8 ones

Note that with whole numbers, you can rely on the number of digits to provide a sense of the relative size of numbers—a 3-digit whole number is always greater than a 2-digit whole number. This is not the case with decimals. When comparing decimals, the number of digits is irrelevant; it is the place value of the digits that matters. Students can use either place value or benchmark numbers to help them compare. For example,

COMPARING USING PLACE VALUE	COMPARING USING A BENCHMARK NUMBER
0.021 < 0.2, since 0 tenths < 2 tenths	0.021 < 0.2, since 0.021 < 0.1 and 0.2 > 0.1
0.021 > 0.01, since 2 hundredths > 1 hundredth	0.021 > 0.01, since 0.021 > 0.02 and 0.01 < 0.02

Many students find it easier to compare (and calculate with) decimals when they have the same number of digits. This is always possible using equivalent decimals. For example,

COMPARING USING EQUIVALENT DECIMALS WITH SAME NUMBER OF DIGITS
0.34 > 0.3, since 0.34 > 0.30 (34 hundredths > 30 hundredths)
8.302 < 8.32, since 8.302 < 8.320 (302 thousandths < 320 thousandths)

Creating In-between Decimals

Sometimes, students need to determine a decimal value between two other values. This is always possible with decimals, as it is with fractions (see Principle 6 on **page 48**). For example,

BETWEEN 3.2 AND 3.6	BETWEEN 3.2 AND 3.3	BETWEEN 3.2 AND 3.21
For decimals between 3.2 and 3.6, you look to the tenths:	For decimals between 3.2 and 3.3, you look to the hundredths:	For decimals between 3.2 and 3.21, you look to the thousandths:
3.2, 3.3, 3.4, 3.5, 3.6	3.2, 3.21, 3.22, ... , 3.29, 3.3	3.2, 3.201, 3.202, ... , 3.209, 3.21

Teaching Idea | **2.49**

Although there is value in using the leading 0 for decimals less than 1 (e.g., 0.43 instead of .43), it sometimes causes difficulty for students, since it is not consistent with what we do with whole numbers (e.g., we don't write 0234 to show that 234 has no thousands).

To focus on the limitations of BID 1, be explicit about this distinction to minimize confusion for students who believe us when we say that decimals work just like whole numbers.

Teaching Idea | **2.50**

Help students see how decimals are like whole numbers (BID 1) by showing how both can be compared using benchmark numbers, by asking: *How do you know 23 < 312?* [23 is less since it is less than 100, but 312 is more than 100] *How might you compare 0.12 and 0.8 using a benchmark number?* [e.g., 0.12 < $\frac{1}{2}$ and 0.8 > $\frac{1}{2}$, so 0.12 < 0.8]

To reinforce BID 1, it is valuable to use BID 2. Ask: *To add 3.45 + 1.38, why could I use 345 hundredths + 138 hundredths instead? To subtract 3.12 − 1.78, why could I use 312 hundredths − 178 hundredths instead?* [When you rename the decimals like this, it allows you to add them as though they are whole numbers.]

Many of these algorithms translate well into mental algorithms, depending on the numbers involved.

Notice how the values of the blocks have changed to represent decimals.

The flat is now the whole (1).

The rod is tenths (0.1).

The unit is hundredths (0.01).

Decimal Operations

As students learn to add, subtract, multiply, and divide with decimals, they should be using what they learned about whole numbers.

- Each principle related to whole number operations continues to apply.
- Each algorithm students learned for whole numbers continues to apply. There are virtually no changes to the explanations for the algorithms when dealing with addition and subtraction of decimals rather than whole numbers. The changes required to deal with multiplication and division with decimals relate more to the subtleties of how things are said than to changes in how the procedures are carried out.
- What students learned about estimating whole numbers also applies to decimals.

Adding and Subtracting Decimals

Just as with whole numbers, students have a choice of algorithms for adding and subtracting decimals.

The Traditional Algorithm for Addition

This algorithm follows the same steps as it does for whole numbers.

Step 1 Model both numbers with blocks. (A place value mat is optional.)

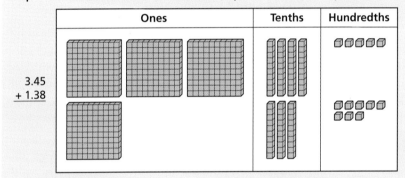

Step 2 Add the hundredths. Trade 10 hundredths for 1 tenth. Record the hundredths that are left.

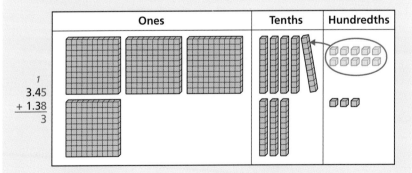

Step 3 Add the tenths. Record the tenths.

Step 4 Add and record the ones.

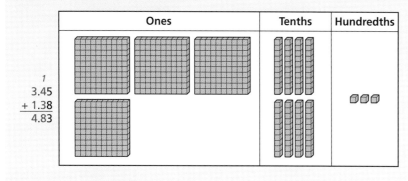

The Traditional Algorithm for Subtraction

As with addition, this subtraction algorithm follows the same steps as it does for whole numbers. It is modelled below using the take away meaning of subtraction.

Step 1 Model the minuend with blocks. (A place value mat is optional.)

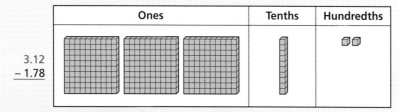

Notice that the subtrahend is not modelled, since it is to be taken away from the minuend.

Step 2A Since there are not enough hundredths to take away 8 hundredths, trade 1 tenth for 10 hundredths. Record the trade above the minuend.

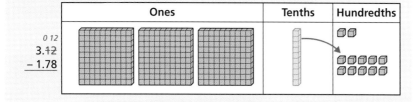

Step 2B Take away 8 hundredths. Record the hundredths that are left.

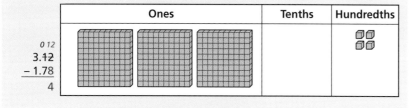

(continued)

Step 3A Since there are not enough tenths to be able to take away 7 tenths, trade 1 one for 10 tenths. Record the trade above the minuend.

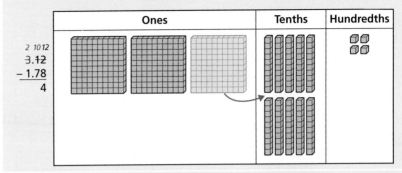

Step 3B Take away 7 tenths. Record the tenths that are left.

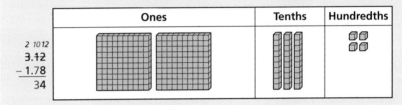

Step 4 Take away 1 one. Record the ones that are left.

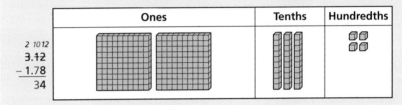

Alternative Algorithms

All of the other algorithms available to students for adding and subtracting whole numbers can be applied to those operations with decimals. Here are a few examples:

"MAKING CHANGE" TO SUBTRACT

Step 1 Think of the subtraction as a missing addend number sentence:

50 − 22.8 = ■, so 22.8 + ■ = 50

Step 2 Count up:

Add 0.2 to get to the next whole number:	22.8 + <u>0.2</u> = 23
Add 7 to get to a multiple of 10:	23 + <u>7</u> = 30
Add 20 to get to the minuend, or total:	30 + <u>20</u> = 50
Add up the "change":	<u>0.2</u> + <u>7</u> + <u>20</u> = 27.2

Often students are taught that to add or subtract decimals, the critical thing to remember is to line up the decimal points. Notice that this is not an issue with many alternative algorithms.

ADDING TOO MUCH AND THEN SUBTRACTING TO COMPENSATE

$$4.25 + 3.97 \rightarrow \quad 4.25 + (3.97 + 0.03) \text{ (Add 0.03 too many.)}$$
$$= 4.25 + 4$$
$$= 8.25 \rightarrow 8.25 - 0.03 \text{ (Take away the extra 0.03.)}$$
$$= 8.22$$

ADDING IN PARTS	SUBTRACTING IN PARTS
5.68 + 3.2	8.46 − 3.7
Start with 5.68 and add 3.2 in parts:	Start with 8.46 and subtract 3.7 in parts:
First add 3: 5.68 + 3 = 8.68	First subtract 3: 8.46 − 3 = 5.46
Then add 0.2: 8.68 + 0.2 = 8.88	Then subtract 0.4: 5.46 − 0.4 = 5.06
	Then subtract 0.3: 5.06 − 0.3 = 4.76

Multiplying and Dividing Decimals

Many of the multiplication and division algorithms for whole numbers can also be applied to decimals, often with the same models.

Multiplying by Powers of 10

Before students begin multiplying with decimals, it is useful to look at patterns with respect to the effect of multiplying whole numbers by 0.1 and 0.01, and decimals by 10, 100, and 1000.

MULTIPLYING WHOLE NUMBERS BY DECIMAL POWERS OF 10 PATTERNS	
$100 \times 400 = 40\ 000$	
$10 \times 400 = 4000$	(One factor is $\frac{1}{10}$ as much, so the product is $\frac{1}{10}$ as much.)
$1 \times 400 = 400$	(One factor is $\frac{1}{10}$ as much, so the product is $\frac{1}{10}$ as much again.)
$0.1 \times 400 = 40$	(One factor is $\frac{1}{10}$ as much, so the product is $\frac{1}{10}$ as much again.)
$0.01 \times 400 = 4$	(One factor is $\frac{1}{10}$ as much, so the product is $\frac{1}{10}$ as much again.)

MULTIPLYING DECIMALS BY WHOLE NUMBER POWERS OF 10 PATTERNS	
$1 \times 2.5 = 2.5$	(One factor is 10 times as much, so the product is 10 times as much.)
$10 \times 2.5 = 25$	(One factor is 10 times as much, so the product is 10 times as much again.)
$100 \times 2.5 = 250$	(One factor is 10 times as much, so the product is 10 times as much again.)
$1000 \times 2.5 = 2500$	(One factor is 10 times as much, so the product is 10 times as much again.)

Moving Digits, Not the Decimal Note that multiplying or dividing by powers of 10 does not change the digits of a number, only the position of each digit within the number. If, for example, you begin with 3.4, dividing by 10 or multiplying by 0.1 decreases the value of each part of the number by a factor of 10, and so the digits in the product change value and move over one place to the right: 3 ones is now 3 tenths and 4 tenths is now 4 hundredths.

3.42

Tens	Ones	Tenths	Hundredths	Thousandths
	3	4	2	

3.42 × 10 = 34.2 or 3.42 ÷ 0.1 = 34.2

Tens	Ones	Tenths	Hundredths	Thousandths
3	4	2		

3.42 ÷ 10 = 0.342 or 3.42 × 0.1 = 0.342

Tens	Ones	Tenths	Hundredths	Thousandths
		3	4	2

When you multiply or divide by powers of 10, it is the digits that move, not the decimal.

Some students are confused by the area model. In the diagram on the right, a student might think that the length of the rectangle is 0.5 instead of 5 because the top row is 5 tenths (i.e., 5 rods) long.

Provide another interpretation (to focus on BID 2, recognizing that a different representation might be more meaningful) by saying that the model also shows 5 sets of 2.3, which is 5 × 2.3. Ask: *How does this model show 5 sets of 2.3?* [There are 5 "columns" with 2 ones (flats) and 3 tenths (rods) in each.] *What would an area model for 3 × 1.7 look like?* [3 "columns" with 1 one (flat) and 7 tenths (rods) in each.]

Many of these algorithms translate well into mental algorithms, depending on the numbers involved.

Using an Area Model to Multiply

This algorithm is similar to the area model used to explain Algorithm 2 for whole numbers (see **page 36**). As with whole-number multiplication, this algorithm can also be modelled using base ten blocks or pictorially.

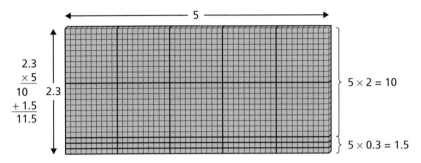

Multiplying by "Ignoring" the Decimal

Often students pretend that the decimal is not there, calculating as if the numbers are whole numbers, and then compensate at the end. This works because the digits do not change; only their placement within the number changes. It is important, however, to estimate to make sure the decimal has been placed correctly.

IGNORE THE DECIMAL (× 10)	COMPENSATE (÷ 10)	ESTIMATE TO CHECK

5 × 2.3 → 5 × 23
   = 5 × 20 + 5 × 3
   = 100 + 15
   = 115
	115 ÷ 10 = 11.5	5 × 2.3 is a bit more than 5 × 2 = 10, and 11.5 is a bit more than 10.
		So, 5 × 2.3 = 11.5.

Multiplying Two Decimals

When multiplying two decimals rather than a decimal and a whole number, students can use the same approaches as shown for whole number and decimals, but they must interpret what the decimals mean. For example, 0.2×0.4 means $\frac{2}{10}$ of $\frac{4}{10}$. Since 0.1×0.4 is $\frac{1}{10}$ of $\frac{4}{10}$, or $\frac{4}{100}$ ($\frac{1}{10} \times \frac{4}{10} = \frac{4}{100}$), then 0.2×0.4 must be twice as much, or 0.08. The grid model below provides a visual representation of what 0.2×0.4 means.

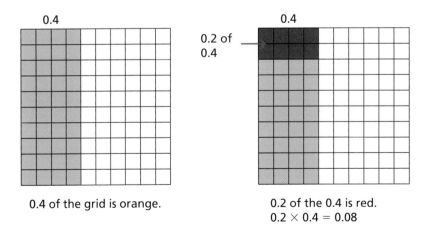

0.4 of the grid is orange.

0.2 of the 0.4 is red.
0.2 × 0.4 = 0.08

Division Algorithms 1 and 2

The algorithms shown below follow the same steps as they do for whole numbers (see Algorithms 1 and 2 for division of whole numbers on **pages 38 to 40**). Because some students experience difficulty lining up the partial quotients correctly using the algorithm on the right, many teachers prefer the one on the left when dealing with decimals.

DIVISION ALGORITHM 1	DIVISION ALGORITHM 2
$$\begin{array}{r} 2.6 \\ 6\overline{)15.6} \\ -\ 12.0 \\ \hline 3.6 \\ -\ 3.6 \\ \hline 0 \end{array}$$	$$\begin{array}{r} 6\overline{)15.6} \\ -\ 12.0 \\ \hline 3.6 \\ -\ 3.6 \\ \hline 0 \end{array}$$ 2.0 groups of 6 0.6 groups of 6 2.6

Mental Math and Decimals

There are many ways to multiply and divide decimals mentally that call upon place value concepts and multiplication and division principles.

Using Place Value Concepts

MULTIPLICATION	DIVISION
$3 \times 0.6 = 3 \times 6$ tenths $= 18$ tenths $= 1.8$	$3 \div 0.6 = 30$ tenths $\div$ 6 tenths $= 5$
$0.3 \times 0.4 = 3$ tenths $\times$ 4 tenths $= 12$ hundredths $= 0.12$	$3.2 \div 0.08 = 32$ tenths $\div$ 8 hundredths $= 320$ hundredths $\div$ 8 hundredths $= 40$

Teaching Idea | **2.53**

Show students how thinking differently can simplify a calculation (BID 1 and 2) by asking: *How could thinking about quarters help you calculate 13.25 ÷ 0.25?* [It's like finding how many quarters are in $13.25, which I could do mentally.] *How could thinking about dimes help you calculate 14.40 ÷ 0.1?* [It's like finding how many dimes are in $14.20, which I could do mentally.]

Using Multiplication and Division Principles

MULTIPLICATION	DIVISION
Double one factor and halve the other. $0.5 \times 6.24 = 1 \times 3.12$ $\qquad\qquad = 3.12$	Increase both numbers by the same factor. $3.0 \div 0.6 = (3.0 \times 10) \div (0.6 \times 10)$ $\qquad\qquad = 30 \div 6$ $\qquad\qquad = 5$
Some students find the relationship between decimals and fractions useful for mental math, particularly for certain decimals that have unit fraction equivalents. For example: $4.28 \times 0.25 = 4.28 \times \dfrac{1}{4}$ $\qquad\qquad = 4.28 \div 4$ $\qquad\qquad = 1.07$	$3.2 \div 0.08 = (3.2 \times 100) \div (0.08 \times 100)$ $\qquad\qquad = 320 \div 8$ $\qquad\qquad = 40$

Proportional Thinking: Rate, Ratio, and Percent

Proportional reasoning focuses on how two amounts are related multiplicatively. For example, thinking of 6 as two 3s instead of as 4 + 2 contrasts multiplicative thinking to additive thinking. Developing proportional reasoning is one of the goals of the Grades 4 to 8 mathematics curriculum. It is only when students can deal with multiplicative relationships that fractions, decimals, ratios, rates, and percents make sense. Proportional reasoning is also essential for students to have a full understanding of other concepts such as measurement with units, the concept of scale (e.g., on maps), geometric notions of similarity, and division. Proportional reasoning also involves qualitative thinking. For example, a student who recognizes that a person who runs 7 km/min finishes a 10-km run slower than one who runs 8 km/min is thinking proportionally.

It is only when students can think proportionally that fractions, decimals, ratios, rates, percents, measurement units, scale, similarity, and division make sense.

BIG IDEAS FOR PROPORTIONAL THINKING

Each teaching idea in this section will indicate which Big Idea(s) for Proportional Thinking (BIPT) can be emphasized.

*Note that many of the big ideas for fractions (**page 42**) also apply to ratios, rates, and percent because ratios, rates, and percent are essentially fractions.*

1. Proportional thinking involves the use of multiplicative relationships, in the form of rates, ratios, and percents, to solve problems.

2. Ratios, rates, and percents, just like fractions and decimals, are comparisons of quantities.

   - A ratio compares quantities with the same unit, for example, 3 boys to 2 girls (the unit being children).

   - A rate compares quantities with different units, for example, distance to time, or price per number of items.

   - A percent always compares a quantity to 100.

Ratio and Rate in the Child's World

Ratio

A ratio is a comparison between two numbers. For example, 3 : 4 is the ratio of blue to red circles shown below. It can be read "three to four" and denotes that for every three blue counters there will be four red counters. The 3 is called the first term of the ratio and the 4 is the second term. Notice that a colon is used between the numbers to represent the ratio.

3 blue to 4 red counters
3 : 4

If a situation can be described by one ratio, it can be described by several ratios. For example, the picture below shows that, for every three boys, there are two girls. Although the picture shows six boys and four girls, they are arranged to emphasize that, for every three boys, there are two girls.

For every 3 boys, there are 2 girls (3 : 2).

All of these ratios describe the situation above:

- 3 : 2 describes boys : girls.
- 3 : 5 describes boys : runners.
- 2 : 5 describes girls : runners.
- 2 : 3 describes girls : boys.
- 5 : 3 describes runners : boys.
- 5 : 2 describes runners : girls.

The ratios 3 : 2 and 2 : 3 are called part-to-part ratios because they compare two parts of something, the boys and the girls. The ratios 3 : 5 and 2 : 5 are called part-to-whole ratios because they compare one part to the whole thing—in this case, the boys or girls to the group of five runners.

Although ratios are often not formally introduced until Grade 6, they are considered much earlier, in informal ways. For example, to teach students about number in kindergarten, you might describe 2 by saying that there are two eyes for every person; this is a ratio. When multiplication is introduced in Grade 2 or 3, ratio is implicit. For example, if you ask students how many wheels are on five bicycles, they use the ratio 2 : 1 to solve the problem.

Rates

Some mathematicians define a rate as like a ratio, only with different units. For example, a rate could be $5.50 per person to represent the cost of a movie, 6 items for $5 to represent the cost of an item, or 12 oranges/litre to represent the number of oranges needed to make 1 L of orange juice. Other mathematicians do not make a distinction between rates and ratios. The notion is that, although the units may be different in a rate, the mathematics used to talk about rates is the same as the mathematics used to talk about ratios. Rates represent comparisons, just as ratios do. Problems involving rates can be solved using the same techniques as problems involving ratios.

Relating Ratio to Fractions

Any ratio can be described as a fraction, for example, 3 oranges to 5 apples, or 3 : 5 as $\frac{3}{5}$. If the ratio is a part-to-whole ratio where the second term describes the whole, the fraction tells what fraction of the whole group the part represents: $\frac{3}{5}$ of the set of fruit is oranges. But even expressing the part-to-part ratio 3 boys to 2 girls, 3 : 2 as the fraction $\frac{3}{2}$ is meaningful. It says that there are $\frac{3}{2}$ as many boys as girls.

Teaching Idea | **2.54**

A good early ratio activity is based on sharing. Show students 25 cookies and say, "Share these cookies in this way—each time Tom gets 2, Kyla gets 3." The intent is to physically model what the ratio 2 : 3 means. Ask students to predict how many cookies each person will get, and then test to see.

Focus on the comparison aspect of a ratio (BIPT 2) by asking: *How did you know that Kyla would get more than Tom?* [Each time, Tom got 2 and Kyla got 3, so she always had more.] *How did you know it would be greater than one more?* [After the second time they took cookies, Kyla had two more and I knew that amount extra would grow.]

Teaching Idea | **2.55**

One of the most "famous" ratios is π, the ratio of circumference to diameter for any circle. The ratio could be written 3.14 : 1 or π : 1 and represents how the circumference compares to the diameter (BIPT 2). Have students measure circular objects of different sizes and notice that, in each case, the result is the same value—slightly greater than 3. Ask: *What other measurement situations have constant ratios?* [e.g., perimeter of a square to its side length is 4 : 1]

The mathematical language used to talk about rates is the same as the mathematical language used for ratios.

Equivalent Ratios and Rates

Two ratios or rates are equivalent if they represent the same relationship. For example, 3 : 4 and 6 : 8 represent the same relationship (as shown below), since if there are 6 blue counters for every 8 red ones, there would have to be 3 blue ones for every 4 red ones for the proportion to stay the same. Similarly, a rate of 6 for $5 is equivalent to a rate of 3 for $2.50.

×	0	1	2	3	4	5	6
0	0	0	0	0	0	0	0
1	0	1	2	3	4	5	6
2	0	2	4	6	8	10	12
3	0	3	6	9	12	15	18
4	0	4	8	12	16	20	24
5	0	5	10	15	20	25	30
6	0	6	12	18	24	30	36

Focus on the comparisons inherent in ratios (BIPT 2) by asking: *Why does it make sense that there are ratios equivalent to 3 : 4 in those two rows?* [The 3 row describes groups of 3 and the 4 row describes the same number of groups of 4, so the ratio is 3 : 4 in each column.] *Where would you look for ratios equivalent to 2 : 5?* [the columns of the ×2 and ×5 rows] *Where else could you look?* [the rows of the ×2 and ×5 columns]

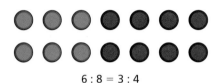

6 : 8 = 3 : 4

Calculating equivalent ratios or rates is handled in the same way as calculating equivalent fractions. To help students make sense of those equivalents, you can use ratio or rate tables. The tables can be either vertical or horizontal. For example, the tables shown below show equivalent ratios for 3 : 4 and equivalent rates for 6 items/$5.

Equivalent Ratios for 3 Blue to 4 Red Counters

If there are this many blue ones,	3	6	9	12	15
there are this many red ones.	4	8	12	16	20

Equivalent Rates for 6 Items for $5

If there are this many items,	6	12	18	24	30
the cost is this much ($).	5	10	15	20	25

Students initially focus on the fact that the values in each column change by adding, but eventually realize that the relationship is really a multiplicative one. The two numbers in each column are the same multiple of the two numbers in any other column.

Using Graphs to Find Equivalent Ratios or Rates

Students can also use graphs to represent ratios or rates and find equivalent ratios or rates. For example, the graph below represents the ratio 1 : 3, comparing the number of cans of frozen juice concentrate to cans of water to make orange juice. The graph can be used to write equivalent ratios, as each ordered pair representing a point on the line describes terms of a ratio equivalent to 1 : 3.

Have students examine a graph like the one on the right to see that 1.5 : 4.5 = 1 : 3. Relate this to the idea that the fraction $\frac{1\frac{1}{2}}{4\frac{1}{2}} = \frac{1}{3}$.

This helps students focus on what is fundamental about $\frac{1}{3}$ (or 1 : 3): that the denominator (or second term) is 3 times (and not 2 more than) the numerator (or first term) (BIPT 1). Ask: *What fraction (or ratio) with a numerator (or first term) of 4.2 would be equivalent to $\frac{1}{3}$?* [$\frac{4.2}{12.6}$]

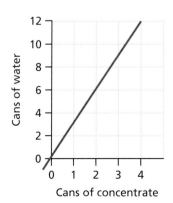

Each ordered pair is an equivalent ratio: (1 : 3), (2 : 6), (3 : 9), ...

Using Ratio or Rate Tables to Find Equivalent Ratios or Rates

Ratio or rate tables list equivalent ratios in an organized way. They can be used to solve problems involving equivalent ratios or rates. For example, the ratio of soccer balls to basketballs in a set of balls is 2 : 5. You can use the ratio table below to solve problems like these:

How many soccer balls are there if there are 15 basketballs?
How many basketballs are there if there are 4 soccer balls?

Soccer Balls	2	?	4
Basketballs	5	15	?

For every 2 soccer balls, there are 5 basketballs.

To find how many soccer balls there are if there are 15 basketballs, notice that $5 \times 3 = 15$ basketballs. That means there are $2 \times 3 = 6$ soccer balls.

Soccer Balls	2	?	4
Basketballs	5	15	?

×3

×3

Soccer Balls	2	6	4
Basketballs	5	15	?

To find how many basketballs there are if there are 4 soccer balls, notice that $2 \times 2 = 4$. That means there are $5 \times 2 = 10$ basketballs.

×2

Soccer Balls	2	6	4
Basketballs	5	15	?

Soccer Balls	2	6	4
Basketballs	5	15	10

×2

New columns are created by multiplying or dividing corresponding numbers in existing columns by any non-zero value.

Another process that results in a column with an equivalent ratio or rate is adding or subtracting corresponding numbers in existing columns. In the table above, you could create a new column by adding $6 + 4$ soccer balls and $15 + 10$ basketballs to create the ratio 10 : 25. Since the ratio 6 soccer balls to 15 basketballs represents the same comparison as 4 soccer balls to 10 basketballs, then combining these, 10 (6 + 4) soccer balls : 25 (15 + 10) basketballs, does not change the comparison.

Some students will not always choose the most efficient approach to determine the desired equivalent ratio or rate. This is quite appropriate if students are still working these ideas through. For example, a student might know that 2 cans of a particular product cost $5 and want to know how much 12 cans would cost. Rather than simply multiplying by 6, their table (and thinking) might look like the following.

If 2 cans cost $5, how much do 12 cans cost?

×2   ×2

Number of Cans	2	4	8	12
Cost ($)	5	10	20	30

$4 + 8 = 12$
$10 + 20 = 30$

Multiplying and adding to complete a rate table

Recipes provide a meaningful context for ratio. Have students adjust recipes to serve more or fewer people. Initially, use multiples and factors of the number of servings, like revising a recipe for 3 to serve 6. Later make the problems more complex, like revising a recipe for 6 to serve 8.

Focus on the multiplicative aspect of proportional thinking (BIPT 1) by asking: *There is $\frac{1}{2}$ cup more flour than sugar in a recipe that serves 2. How do you know there will be $1\frac{1}{2}$ cups more flour than sugar in the revised recipe for 6?* [If you triple the recipe, the difference between the amounts will be tripled too.]

Using Unit Rates

There are many ways to solve rate problems; using unit rates is just one strategy.

A *unit rate* is an equivalent rate where the second term is 1. For example, if you drive 30 km in 20 min, the unit rate is 1.5 km/min (the second term is 1 min). Unit rates are just one way to solve rate problems. Here is how you would solve the following problem using unit rates:

Which is the better buy: 3.6 L of dish-washing soap A for $3.69 or 4 L of dish-washing soap B for $4.29?

Teaching Idea | **2.59**

Use rate problems (BIPT 1) that are of interest to students like this:

Hair grows at a rate of about 10 cm per year. About how long does it take to grow 4 cm? [about 146 days, or almost 5 months]

Help students see that there are many strategies involving multiplicative relationships for solving the problem, such as finding an equivalent fraction to $\frac{4}{10}$ with a denominator of 365 (by multiplying the numerator of 4 by 36.5), or using a ratio table.

USING A LITRES/DOLLAR UNIT RATE TO SOLVE THE PROBLEM	
If you think of the rate as litres/dollar, you will need to know how many litres of each soap you can buy for $1. If you get more litres of one soap than another for $1, it is the better buy.	If 3.6 L of A cost $3.69, you need to divide 3.6 by 3.69 to see how much you get for $1: 3.6 ÷ 3.69 = 0.9756 L → 0.9756 L/dollar. If 4 L of B cost $4.29, you divide 4 ÷ 4.29 to see how much you get for $1: 4 ÷ 4.29 = 0.9324 L → 0.9324 L/dollar. Soap A is the better buy: 0.9756 L > 0.9324 L, so you get more for each $1.

USING A DOLLARS/LITRE UNIT RATE TO SOLVE THE PROBLEM	
If you think of the rate as dollars/litre, you will need to know how many dollars 1 L of each soap costs. If it costs fewer dollars for 1 L of one soap than another, it is the better buy.	If $3.69 gets you 3.6 L of A, then 1 L costs 3.69 ÷ 3.6 = $1.025 ≈ $1.03. If $4.29 gets you 4 L of B, then 1 L costs 4.29 ÷ 4 = $1.0725 ≈ $1.07. Soap A is the better buy: $1.03 < $1.07, so it costs less for each litre.

You could also solve the problem above using two rate tables:

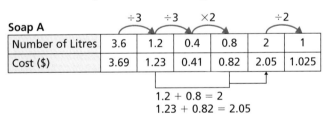

Soap A

	÷3	÷3	×2		÷2
Number of Litres	3.6	1.2	0.4	0.8	2	1
Cost ($)	3.69	1.23	0.41	0.82	2.05	1.025

1.2 + 0.8 = 2
1.23 + 0.82 = 2.05

Soap B

	÷2	÷2
Number of Litres	4	2	1
Cost ($)	4.29	2.145	1.0725

Relating and Comparing Ratios and Rates

As with fractions, we can easily compare ratios or rates with the same second term. For example, if the part-to-whole ratio of skiers to the population of a school is 72 : 410, and the ratio of snowboarders to the population of that school is 189 : 410, it is easy to see that the proportion of snowboarders to school population is greater than that for skiers. Similarly, if the cost of 5 cans of one vegetable is $4 and the cost of 6 cans of another is $4, it is easy to see that the second type of vegetable costs less.

Part-to-part ratios are often compared to the benchmark ratio 1 : 1 (which represents $\frac{1}{2}$). For example, the ratio of skiers to non-skiers (72 : 338) is less than the benchmark ratio 1 : 1 (205 : 205).

Solving Ratio and Rate Problems

Ratios or rates can be used to solve a variety of problems. Some teachers begin using problems that do not involve numbers. The idea is for students to get a sense of what the situations actually mean. For example, students are presented with these two glasses of liquid and are asked to predict which will taste sweeter.

Which Will Taste Sweeter?

| 1 PART ORANGE JUICE TO 2 PARTS WATER | 1 PART ORANGE JUICE TO 1 PART WATER |

Some educators use a problem in which students compare parking lots to determine which is more full (as shown below). Students cannot just count cars since the lots are of different sizes; they must use proportional thinking. Initially, they might just have a feeling that Lot B looks more full. Then they might figure out that there are 24 cars in 40 spaces on the left, and 56 cars in 80 spaces on the right. 56 cars in 80 spaces is more full since it is equivalent to 28 cars in 40 spaces, and 28 is greater than 24.

Teaching Idea | **2.60**

Ask students to enlarge this picture so it is twice as wide and twice as high.

To focus on proportional thinking (BIPT 1), ask: *Why is this a ratio problem?* [You can multiply to compare the width or length of the small picture to the larger one and you're using the same units, i.e., 4 units tall to 8 units tall.]

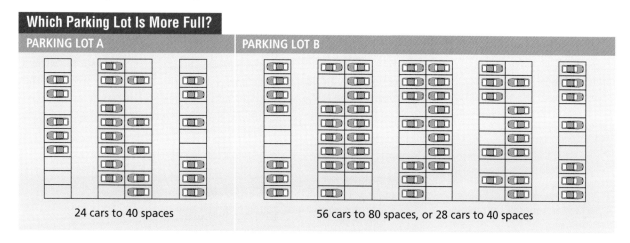

Which Parking Lot Is More Full?

| PARKING LOT A | PARKING LOT B |

24 cars to 40 spaces   56 cars to 80 spaces, or 28 cars to 40 spaces

It is important for students to understand that there is no one correct method for solving ratio problems. For example, if the ratio of boys to girls in a class is 5 : 4 and you want to determine the possible number of students there could be in the class, students might keep drawing diagrams of 5 boys and 4 girls until an appropriate total class size is achieved. Or, they might add 5 + 4 = 9 to represent a group of girls and boys in the correct ratio, and then use appropriate multiples of 9 as the class size: 18, 27, and 36.

Teaching Idea | **2.61**

Present a variety of ratio and rate problems:

- After 2 h, 5 h, and 7 h of driving, 260 km, 650 km, and 890 km were travelled. Did the three drivers travel at the same rate? [no]

- Casey spent $37.38 for 42 L of gas. At that rate, could she get another 40 L for less than $35? [no]

- A teacher divided 25 students into 5 equal groups, each with 3 girls. How many boys are in the class? [10]

- 7 girls are sharing 3 cheese pizzas, and 3 boys are sharing 1 pepperoni pizza. Who gets more pizza, a boy or a girl? [a girl, if the pizzas are the same size]

- Is an 8 × 10 enlargement of a 5 × 7 picture an exact enlargement of the original picture? [no]

- There are about 105 males born in Canada for every 100 females. In Calgary, there were about 15 000 births in 2007. About how many of these babies were boys? [about 7700]

To focus on proportional thinking (BIPT 1), ask: *How are all of these problems similar?* [they all involve comparisons where you multiply or divide]

A percent is a special ratio with a second term of 100.

These principles can be explored using a variety of models and approaches.

Solving Proportions

Later, students solve quantitative problems, which are often solved using proportions. In this case, it is important for students to understand that a proportion is an expression of the equivalence of two ratios. For example, if $\frac{3}{5} = \frac{x}{20}$, then the ratio $3 : 5$ is the same as the ratio $x : 20$.

There are many ways to solve for x, including looking for a possible multiplier for both terms of the ratio $3 : 5$ (in this case 4), or perhaps using a ratio table or a diagram.

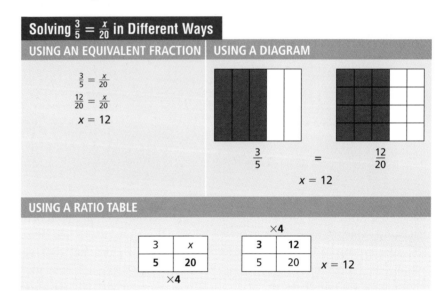

Percent

Percents are a special sort of ratio—a ratio where the second term is 100. The concept of 50% or 100% is rarely new to students. They are familiar with both from talk about test scores and everyday situations such as weather reports. In fact, some teachers choose to teach percent before they teach other ratios. Some students may note the connection to the word "cent," where a cent is $\frac{1}{100}$ of a dollar, just like 1 cm is $\frac{1}{100}$ of 1 m.

Percent Principles

PERCENT PRINCIPLES

1. A percent is a ratio or a comparison of a number to 100 and can be written as ■ : 100, or as $\frac{■}{100}$, or as an equivalent decimal. It can be represented with a variety of models.

2. The actual amount that a percent represents is based on the whole of which it is a percent.

3. Comparing percents is as easy as comparing decimals, since in many instances you only need to compare whole-number values.

4. Percents can be as low as 0 and can go higher than 100.

5. Sometimes percents are used to describe change.

6. There are a variety of appropriate strategies for solving any problem involving percent.

PRINCIPLE 1 A percent is a ratio or a comparison of a number to 100 and can be written as ■ : 100, or as $\frac{■}{100}$, or as an equivalent decimal. It can be represented with a variety of models.

Early on, any model for a percent should involve something that has 100 parts; this will reinforce the meaning of percent as a comparison to 100.

Percent Models

BASE TEN BLOCKS	A DECIMAL HUNDREDTHS GRID
One obvious model is a base ten block flat as the whole. To show 22%, you can cover a base ten flat with 2 rods and 2 units.	A decimal hundredths grid is an excellent model for percents, particularly if grids have been used previously to model fractions and decimal hundredths.

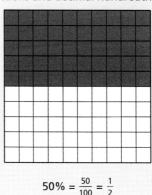

22% of the flat is covered.

$$50\% = \frac{50}{100} = \frac{1}{2}$$

A percent can always be written as a decimal, or vice versa. For example, 48% is the same as 0.48; both mean 48 hundredths. If a number is written as a decimal, the first two places to the right of the decimal point can be written as a whole number percent. For example, 0.235 can be written as 23.5%.

Initially, students should relate certain percents to some comfortable fractions, for example, 50% to $\frac{1}{2}$, 25% to $\frac{1}{4}$, 10% to $\frac{1}{10}$, and perhaps 33% to about $\frac{1}{3}$. Later, students use relationships to relate a broader group of percents to fractions or fractions to percents. For example, the fraction $\frac{3}{4}$ is related to 75% since it is the same as three groups of $\frac{1}{4}$, or $3 \times 25\%$.

PRINCIPLE 2 The actual amount that a percent represents is based on the whole of which it is a percent.

Since a percent is essentially a fraction with a denominator of 100, the key fraction concept of the importance of knowing what the whole is also applies to percent. 1% can be a lot or a little; it depends what it is 1% of. For example, 1% of all the water on the planet is a lot of water, but 1% of a glass of water is not very much water at all.

Also, the same quantity can represent different percents depending on the whole. For example, 20 is 50% of 40, but it is 100% of 20 and only 10% of 200.

PRINCIPLE 3 Comparing whole-number percents is as easy as comparing two decimals or two whole numbers.

Although comparing two fractions with different numerators and denominators is not always straightforward, comparing two percents is as easy as comparing whole numbers or decimals. For example, it is not immediately obvious that the fraction $\frac{27}{40}$ is greater than the fraction $\frac{21}{32}$. However, if they are both written as decimals (0.675 and 0.65625) or percents (67.5% and 65.625%), it is clear that $\frac{27}{40} > \frac{21}{32}$.

Teaching Idea | **2.62**

We often ask students questions such as "What is 10% of 320?" The single correct reply is 32.

Try going backwards for a richer task that focuses on BIPT 1 and 2, by asking students to complete 32 is ___% of ____ in as many ways as they can. [32 is 100% of 32, 32 is 50% of 64, 32 is 25% of 128, ...]

Teaching Idea | **2.63**

Focus on BIPT 1 and 2 by helping students see that it is not possible to interpret a percent meaningfully without knowing what the whole is. Ask: *I painted 50% of one picture red and 20% of another picture red. Which picture used more paint?* [It depends on the size of each; if they are the same size, then the first picture used more.]

PRINCIPLE 4 Percents can be as low as 0, but can go higher than 100.

Fraction and decimal percents less than 1 can be compared to the whole. For example, 0.5% means half of 1%. It actually represents the fraction $\frac{1}{200}$, since there would be 200 half-squares and only 1 is coloured.

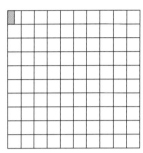

0.5% is half of 1%.
0.5% is $\frac{1}{200}$ of the grid.

Teaching Idea | **2.64**

Using a diagram such as the one on the right, ask students why someone might describe this as 62.5% instead of 125%. [the whole, or 100%, is 2 grids instead of 1]

To encourage flexibility in thinking and to focus on BIPT 2, ask: *What other percent could this represent? What would 100% look like each time?* [e.g., 41.7% if 3 grids are 100%]

To show a percent more than 100%, it is necessary to use more than one grid, assuming each grid is 100%. For example,

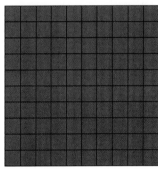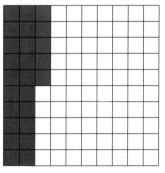

125%

Using more than one grid to model a percent greater than 100%

The ability to represent a percent increase or decrease has many practical, everyday applications.

PRINCIPLE 5 Sometimes percents are used to describe change.

A newspaper headline might read, "The population grew by 8%." One of the important ideas for students to understand is that the 8% represents the change and the new population is 108% of the old one. The ability to represent a percent increase (or decrease) has many practical, everyday applications. For example, if 13% tax has to be added to the price of an item, you might be interested in knowing what the new price is as a percent of the original price (113%).

Sometimes prices are decreased. For example, a store is having a 40% off sale. Knowing that the new price is 60% of the original price allows for a quick estimate of the new price.

Another situation where percents are used to describe change is currency conversion.

Currency Conversions

If the rate of exchange is C$1 = US$0.92, you can use percents to figure out the appropriate conversion.

If the Canadian dollar is 92% of the U.S. dollar, a dollar amount in Canadian dollars must be multiplied by 0.92 to determine its worth in U.S. dollars.	If the Canadian dollar is 92% of the U.S. dollar, a U.S. dollar amount must be divided by 0.92 to determine its worth in Canadian dollars.
C$25 = US$25 × 0.92 = US$23	US$23 = C$23 ÷ 0.92 = C$25

If C\$1 is worth US\$0.92 (a loss of 8¢), it turns out that US\$1 is worth C\$1.09 at the same exchange rate (a gain of 9¢). Many people are surprised that you do not lose and gain the same number of cents when you make a currency conversion. The reason is based on Principle 2. The whole that you are taking a percent of has changed; in one case, the whole is the Canadian dollar, and in the other case, it is the U.S. dollar, which is worth more, making it a greater whole.

PRINCIPLE 6 There are a variety of appropriate strategies for solving any problem involving percent.

Sometimes, you want to calculate the percent; other times, you want to calculate the amount the percent represents; and still other times you want the value of the whole.

Solving a Percent Problem (Discount Amount) in Different Ways

Problem 1 The price of a \$60 item has been reduced by 20%. How much will you save?

To calculate 20% of \$60, you can write 20% as the decimal 0.2 and multiply by 60.

$0.2 \times 60 = 12$

You save \$12.

To calculate 20% of \$60, you can think of 20% as $\frac{1}{5}$ and divide 60 by 5 or multiply 60 by $\frac{1}{5}$.

$60 \div 5 = 12$ or $60 \times \frac{1}{5} = 12$

You save \$12.

To calculate 20% of \$60, you can use a ratio table.

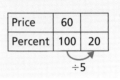

Price	60	
Percent	100	20

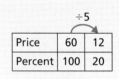

Price	60	12
Percent	100	20

You save \$12.

Solving a Percent Problem (Discount Percent) in Different Ways

Problem 2 You paid \$50 for an \$80 item. What was the discount percent?

To calculate what percent \$30 is of \$80, you can compare 30 and 80 as the fraction $\frac{30}{80}$, and then calculate the equivalent decimal.

$\frac{30}{80} = 0.375$

$0.375 = 37.5\%$

The percent discount was 37.5%.

To calculate what percent \$30 is of \$80, you might realize that an \$8 savings is a 10% discount. But you saved \$30.

Since $30 \div 8 = 3\frac{3}{4}$, the percent you want is $10\% \times 3\frac{3}{4} = 37.5\%$.

The percent discount was 37.5%.

To calculate what percent \$30 is of \$80, you can solve this proportion:

$$\frac{30}{80} = \frac{x}{100}$$

To calculate x, you can multiply:

$\frac{30}{80} \times 100 = 37.5$

The percent discount was 37.5%.

Solving a Percent Problem (Original Amount) in Different Ways

Problem 3 You saved \$50 and the discount was 40%. What was the original price?

You might reason that, if a discount of 40% is \$50, a discount of 20% is \$25. If 20% is \$25, then the original price (100%) must be 5 times as much, or \$125.
The original price was \$125.

You can divide \$50 by 0.4, since the original price was multiplied by 0.4 to get \$50. The result is \$125.
The original price was \$125.

You can use a ratio table.

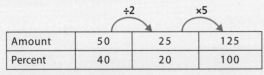

Amount	50	25	125
Percent	40	20	100

The original price was \$125.

You can solve a proportion: $\frac{40}{100} = \frac{50}{x}$

To calculate x, you can multiply both sides of the equation by $100x$, and then divide by 40. The result is \$125.
The original price was \$125.

Integers

The concept of negative integers is introduced near the end of the elementary school years or in early middle school. Many students, even younger ones, are quite comfortable interpreting these numbers, particularly because of their familiarity with negative temperatures, although operations with integers will be less familiar.

BIG IDEAS FOR INTEGERS

Each Teaching Idea in this section will indicate which Big Idea(s) for Integers (BII) can be emphasized.

1. The negative integers are the "opposites" of the whole numbers. Each integer is the reflection of its opposite across a line that is perpendicular to the number line at 0. For example,

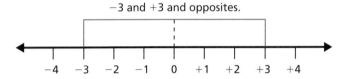

−3 and +3 and opposites.

2. In a number of ways, integers are more like whole numbers than like fractions or decimals.

3. The zero property, that is, $(-1) + (+1) = 0$, plays an important role in many integer operations.

4. The meanings for the operations that apply to whole numbers, fractions, and decimals also apply to all integers. Each meaning can be represented by a model, although some models suit some meanings better than others.

Integers in the Child's World

Children may be familiar with "minus" numbers, but they may not interpret them mathematically, as the opposites of the positive integers. In fact, what makes −1 negative 1 is that it is just as far from 0 as its opposite, +1, is from 0. This idea can be best understood by students using a number line, vertical or horizontal.

Integer Contexts

Useful contexts for making work with integers meaningful, depending on the interest and experience of students, include

- temperatures
- floors below and above a main floor
- being in debt or not
- below and above sea level or ground level
- golf scores that are below and above par
- hockey for and against scores

Teaching Idea | **2.65**

To help students think of opposite integers as reflections across 0 (BII 1), have them hold a transparent mirror perpendicular to the number line at 0 and see where the image of each integer appears. Ask: *The image of −2 is at +2. What does that tell you about −2 and +2?* [They are an equal distance from 0, so they are opposites.]

Reading and Writing Integers

Some people use a raised $+$ or $-$ sign to make the distinction between the integer signs and the operations signs for addition and subtraction. However, even though making this distinction might be helpful when students first work with integers, later on they will learn that there is a connection between subtraction and the negative sign and between addition and the positive sign; for example, $-3 = 0 - 3$ and $+3 = 0 + 3$. The preference might be to use the same signs for integers as for operations but put brackets around each integer to separate the positive and negative signs from the operation signs; for example, $(+3) - (-4) + (+2) = (+9)$. The brackets can be dropped when there is no operation sign in front of the integer; for example, $+3 - (-4) + (+2) = +9$.

When students first work with integers, they use the $+$ sign to indicate a positive integer, even though it is not necessary to do so, as the sign is assumed if there is no integer sign; for example, $(+3) - (-4) + (+2) = (+9)$ is the same as $3 - (-4) + 2 = 9$. The $+$ sign can be dropped as soon as students are beyond the introductory stage of integers.

Comparing Integers

When comparing integers, it is helpful if students think about the placement of each integer on a number line rather than trying to memorize rules. In this way, students are much less likely to make the mistake of thinking that -9 is less than -7.

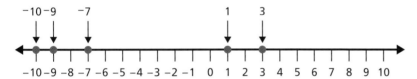

When using a number line model, it is obvious that $-10 < -9 < -7 < 1 < 3$.

Principles for Comparing Integers

PRINCIPLES FOR COMPARING INTEGERS

1. Any negative integer is always less than any positive integer.
2. A positive integer closer to 0 is always less than a positive integer farther away from 0; for example, $+1 < +3$.
3. A negative integer closer to 0 is always greater than a negative integer farther away from 0; for example, $-7 > -10$.

PRINCIPLE 1 Any negative integer is always less than any positive integer.
Every negative value is to the left of 0. Every positive value is to the right of 0. Since the number line is built so that greater numbers are to the right, any positive integer must be greater than any negative integer.

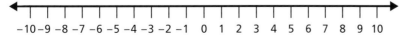

All negative integers are to the left of all positive integers, so any negative integer is less than any positive integer.

Teaching Idea | **2.66**

To focus on how integers are more like whole numbers than fractions (BII 2), ask: *How many whole numbers are between 2 and 10?* [7] *How many fractions or decimals are between 2 and 10?* [an unlimited number] *How many integers are between -2 and -10?* [7] *Why are integers more like whole numbers than fractions and decimals?* [There are no integers (or whole numbers) between consecutive integers (or whole numbers), but there is always a fraction (or decimal) between any two fractions (or decimals).]

These principles can all be explained using a number line model.

PRINCIPLE 2 A positive integer closer to 0 is always less than a positive integer farther away from 0.

Zero is to the left of all of the positive integers. So a positive integer closer to 0 is to the left of one farther from 0. Since it is farther to the left, it is less; for example, $+4 < +8$.

PRINCIPLE 3 A negative integer closer to 0 is always greater than a negative integer farther away from 0.

Zero is to the right of all of the negative integers. So a negative integer closer to 0 is to the right of, and therefore greater than, a negative integer farther from 0; for example, $-6 > -9$.

The Zero Principle

Mathematicians have defined (-1) as the number that you add to $+1$ to result in 0; that is, by definition, $(-1) + (+1) = 0$. This is referred to as the zero principle and is the foundation for many computations involving negative numbers. As a consequence of this definition, any number can be added to its opposite to result in a value of 0. For example,

$$(-3) = (-1) + (-1) + (-1) \text{ and } (+3) = (+1) + (+1) + (+1)$$
$$\text{Thus, } (-3) + (+3) = (-1) + (-1) + (-1) + (+1) + (+1) + (+1).$$

Since the order of adding numbers is irrelevant,
$$(-3) + (+3) = [(-1) + (+1)] + [(-1) + (+1)] + [(-1) + (+1)]$$
$$= 0 + 0 + 0$$
$$(-3) + (+3) = 0$$

The zero principle is used in situations such as $(-2) - (-5)$, where you may want to take away (-5) but you only have (-2).

Teaching Idea | **2.67**

Have students choose the subtractions they would use the zero principle for (BII 3):

$$-3 - (-1) \qquad -8 - (+3)$$
$$6 - 2 \qquad 10 - (-4)$$

[for $-8 - (+3)$ and $10 - (-4)$, because there is not enough in the minuend to take away the subtrahend]

Adding Integers

Just as with whole numbers, adding means putting together. Therefore, adding $(+3) + (-5)$ means putting together an amount representing $+3$ with an amount representing (-5). This can be modelled in different ways:

Adding $(-5) + (+3)$	
USING A NUMBER LINE	**USING COUNTERS**
Start at (-5) and move forward $(+3)$ to end up at (-2).	Use the zero principle, $(-3) + (+3) = 0$, to simplify $(-5) + (+3)$ to (-2).

$$(-5) + (+3) = (-2)$$

$$(-3) + (+3) = 0$$

$$(-5) + (+3) = (-2)$$

Principles for Adding Integers

It is important for students to have a conceptual understanding of the principles that underpin integer addition.

PRINCIPLES FOR ADDING INTEGERS

1. The sum of two negatives is negative.
2. The sum of two positives is positive.
3. The sum of a negative and a positive can be positive or negative. The sum has the sign of the number that is farther from 0.

PRINCIPLE 1 The sum of two negatives is negative.

By modelling with blue counters to represent negative integers, students see that when they put, for example, 3 blue counters together with 4 blue counters, $(-3) + (-4)$, the result is 7 blue counters, (-7).

PRINCIPLE 2 The sum of two positives is positive.

By modelling with red counters to represent positive integers, students see that when they put, for example, 5 red counters together with 2 red counters, $(+5) + (+2)$, the result is 7 red counters, $(+7)$.

PRINCIPLE 3 The sum of a negative and a positive can be positive or negative. The sum has the sign of the number that is farther from 0 on a number line.

This principle can be modelled using a number line or using counters, as shown below.

ADDING $(-3) + (+4)$	ADDING $(+3) + (-4)$
The sum of $(-3) + (+4)$ is positive since there are more red (positive) counters than blue (negative) counters.	The sum of $(+3) + (-4)$ is negative since there are more blue (negative) counters than red (positive) counters.
1 red counter $(+1)$ is left over when zeros are created by pairing up reds and blues (the zero principle).	1 blue counter (-1) is left over when zeros are created by pairing up reds and blues (the zero principle).

$(-3) + (+4) = (+1)$	$(+3) + (-4) = (-1)$
Symbolically	**Symbolically**
You can break up $(+4)$ into $(+3)$ and $(+1)$ in order to use the zero principle:	You can break up (-4) into (-3) and (-1) in order to use the zero principle:
$(-3) + (+4) = (-3) + (+3) + (+1)$ $= 0 + (+1)$ $= +1$	$(+3) + (-4) = (+3) + (-3) + (-1)$ $= 0 + (-1)$ $= -1$

Teaching Idea | **2.68**

To focus on BII 3, ask: *Why do you have to know the zero principle to add a positive and a negative integer but not to add two positive integers or two negative integers?* [When you add two positives (or negatives), there are no opposites to combine to make zero.]

Teaching Idea | **2.69**

Integrate integer work with probability. Have students toss a coin. If the coin lands heads, they gain a point by adding $(+1)$. If it lands tails, they lose a point by adding (-1). After 20 tosses, they figure out their final score.

To bring out BII 3, ask: *Someone had a final score of $(+2)$. How many tails and heads did they toss?* [9 tails and 11 heads]

Why is it helpful to understand the zero principle to answer this question? [You can think about 9 pairs of tosses, each a head and a tail worth 0, and then 2 more head tosses to end up with $(+2)$.]

Encourage students to use a variety of mental math strategies to add integers; for example:

$$(-32) + (-39) + 42 + 30$$
$$= 42 + (-32) + (-39) + 30$$
$$= 10 + (-9)$$
$$= 1$$

Subtracting Integers

Principles for Subtracting Integers

As with adding integers, it is important for students to have a conceptual understanding of the principles that underpin integer subtraction.

PRINCIPLES FOR SUBTRACTING INTEGERS

1. To subtract an integer, you can use either a take away meaning or a missing addend meaning.
2. To subtract an integer, you can use the zero principle to add enough to the minuend (the starting amount) to take away the subtrahend.
3. To subtract an integer, you can determine what to add to the subtrahend to get to the minuend; sometimes you need to use the zero principle.
4. To subtract an integer, you can add its opposite.

PRINCIPLE 1 To subtract an integer, you can use either a take away meaning or a missing addend meaning.

As with whole numbers and decimals, subtraction can mean take away or how much must be added to one number to get another. You might model subtraction of integers with two-colour counters or a number line using either a take away or a missing addend meaning.

Modelling Subtraction

USING THE TAKE AWAY MEANING	USING THE MISSING ADDEND MEANING
$(-5) - (-2)$ means you have 5 blue counters and you want to take away 2 of them, so 3 blue counters are left.	$(-5) - (-2)$ is the amount that has to be added to (-2) to get to (-5).

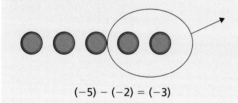

$$(-5) - (-2) = (-3)$$

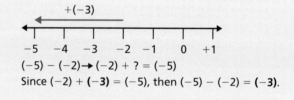

$(-5) - (-2) \rightarrow (-2) + \, ? = (-5)$
Since $(-2) + (-3) = (-5)$, then $(-5) - (-2) = (-3)$.

PRINCIPLE 2 To subtract an integer, you can use the zero principle to add enough to the minuend (the starting amount) to take away the subtrahend.

Both meanings of subtraction can be used to subtract numbers like $(-2) - (-6)$ and $(+2) - (+6)$, where taking away is not immediately possible, since you only have 2 blue or 2 red counters and you need to take away 6 blue counters.

USING THE TAKE AWAY MEANING FOR $(-2) - (-6)$	USING THE MISSING ADDEND MEANING FOR $(+2) - (-6)$
$(-2) - (-6)$ means you have 2 blue counters and want to take away 6 blue counters. To get 6 blue counters to take away, you can add 4 pairs of red and blue counters, each pair with a value of 0, without changing the value of the minuend. Once you have 6 blue counters, you can take them away and there are 4 red counters left, or $(+4)$.	$(+2) - (-6)$ is the value of the counters that you have to add to 6 blue counters (-6) to end up with 2 red counters $(+2)$. If you add 8 red counters to 6 blue counters, 6 red counters pair with the 6 blue counters for a total value of 0, and you end up with 2 red counters $(+2)$.

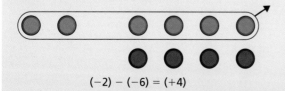

$$(-2) - (-6) = (+4)$$

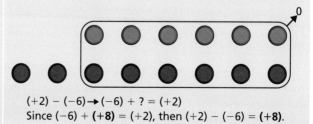

$(+2) - (-6) \rightarrow (-6) + \, ? = (+2)$
Since $(-6) + (+8) = (+2)$, then $(+2) - (-6) = (+8)$.

PRINCIPLE 4 To subtract an integer, you can add its opposite.

Some teachers teach subtraction of integers as the rule, "To subtract an integer, add the opposite instead." For example, to solve $(-3) - (-5)$, they encourage students to rewrite the calculation as $(-3) + (+5)$. Although the correct answer is achieved and students will accept the rule, it is not really clear to most students why the rule works.

When you subtract a negative from a positive using the number line model, it is easy to see why you can add the opposite to subtract.

For example, $(+4) - (-5)$ tells what to add to (-5) to get to $(+4)$. To go from (-5) to $(+4)$, you move 5 to the right $(+5)$ to get to 0, and then another 4 to the right $(+4)$ to get to $(+4)$, so the total amount to be added is $(+5) + (+4)$ or $(+4) + (+5)$.

Teaching Idea | **2.70**

To emphasize BII 4, that one meaning or approach might be better than another, ask: *Why might you calculate $(-40) - (-5)$ using take away instead of using the "add the opposite" rule?* [It's easy to take 5 things away from 40 of the same thing. Adding the opposite would be an extra step.]

$$(+4) - (-5) \rightarrow (-5) + \underline{\hspace{2cm}} = (+4)$$
$$(-5) + \underline{(+5) + (+4)} = (+4)$$

Since $(-5) + \underline{(+5) + (+4)} = (+4)$, then $(+4) - (-5) = \underline{(+4) + (+5)}$.

The concept that subtracting is the same as adding the opposite can also be used for explaining subtracting a negative from a negative, such as $(-3) - (-2)$, but it tends to be a little less obvious. The rule can be justified using equations, as shown below.

USING EQUATIONS TO EXPLAIN WHY THE RULE "ADD THE OPPOSITE" WORKS

$(-3) - (-5) = \blacksquare \rightarrow (-5) + \blacksquare = (-3)$	(Change the subtraction equation to a missing addend equation.)
$(+5) + (-5) + \blacksquare = (-3) + (+5)$	(Add $(+5)$ to both sides.)
$0 + \blacksquare = (-3) + (+5)$	(Simplify.)
$\blacksquare = (-3) + (+5)$	

So, if $(-3) - (-5) = \blacksquare$ and $\blacksquare = (-3) + (+5)$, then $(-3) - (-5) = (-3) + (+5)$.

This explanation is highly symbolic and may be more difficult for most students than using the number line (as described above).

Multiplying Integers

Principles for Multiplying Integers

Although the principles for multiplying integers are fairly easy for students to learn, explaining why those principles make sense is a greater challenge.

PRINCIPLES FOR MULTIPLYING INTEGERS

1. Integers can be multiplied in any order without affecting the product.
2. The product of a positive and a negative is negative.
3. The product of two positives or two negatives is positive.
4. The distributive property applies to multiplication and addition of integers; that is, $a(b + c) = ab + ac$.

PRINCIPLE 1 Integers can be multiplied in any order without affecting the product. Mathematicians have defined the set of integers to ensure that they follow the same principles as the set of whole numbers. One example is the commutative property. For example, $(+3) \times (-2) = (-2) \times (+3)$. Knowing this property is key to explaining Principle 2.

PRINCIPLE 2 The product of a positive and a negative is negative.

A variety of models can be used to explain why a positive multiplied by a negative is negative.

Multiplying a Positive by a Negative

USING A COUNTER MODEL

Since $(+3) \times (-2) = 3 \times (-2)$, you can model $(+3) \times (-2)$ as 3 groups of (-2) counters.

$$(+3) \times (-2) = 3 \times (-2) = (-6)$$

USING A NUMBER LINE MODEL

Since $(+3) \times (-2) = 3 \times (-2)$, you can model $(+3) \times (-2)$ as 3 moves of (-2) on a number line.

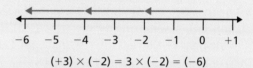

$$(+3) \times (-2) = 3 \times (-2) = (-6)$$

USING REPEATED ADDITION

If you use the repeated addition meaning of multiplication, you will discover that the product is negative:

$$(+3) \times (-2) = 3 \times (-2) = (-2) + (-2) + (-2) = (-6)$$

Multiplying a Negative by a Positive

USING THE COMMUTATIVE PROPERTY

For $(-3) \times (+2)$, since it is not clear what (-3) groups means, you can use the commutative property:

$(-3) \times (+2) = (+2) \times (-3) = 2 \times (-3)$, which can be modelled as 2 groups of 3 blue counters or 2 moves of (-3) on a number line.

USING A PATTERN

$(+3) \times (+2) = (+6)$ $(+2) \times (+2) = (+4)$ $(+1) \times (+2) = (+2)$ $0 \times (+2) = 0$ $(-1) \times (+2) = ?$ $(-2) \times (+2) = ?$ $(-3) \times (+2) = ?$	The product decreases by 2 each time, as the first factor decreases by 1, which makes sense since each time there is 1 fewer group of 2. To continue the pattern, the products must be negative: $(-1) \times (+2) = (-2)$ $(-2) \times (+2) = (-4)$ $(-3) \times (+2) = (-6)$

PRINCIPLE 3 The product of two positives or two negatives is positive.

It is easy to understand why the product of two positives is positive by using counters or a number line or by relating it to what students know about whole numbers.

One of the most comfortable ways for students to understand why the product of two negatives is positive, for example, $(-3) \times (-2) = (+6)$, is to use patterns. You can set up a pattern that will ultimately lead to the right conclusion.

Teaching Idea | **2.71**

The zero principle is clearly fundamental to integer addition and subtraction, but its role in multiplication and division is subtle. Its role in multiplication (BII 3) could be pointed out by asking: *Look at the pattern to the right. Why is it important to understand the zero principle to be able to create the pattern?* [You use the zero principle to continue the pattern in the products by adding $(+2)$ each time.]

USING A PATTERN TO EXPLAIN WHY NEGATIVE × NEGATIVE = POSITIVE

$(+3) \times (-2) = (-6)$ $(+2) \times (-2) = (-4)$ $(+1) \times (-2) = (-2)$ $0 \times (-2) = 0$ $(-1) \times (-2) = ?$ $(-2) \times (-2) = ?$ $(-3) \times (-2) = ?$	As the first factor decreases by 1, the product increases by 2, which makes sense since, each time there is 1 fewer group of (-2) to decrease the product. To continue the pattern, the products must be positive: $(-1) \times (-2) = (+2)$ $(-2) \times (-2) = (+4)$ $(-3) \times (-2) = (+6)$

Some teachers use "stories" to explain why the product of two negatives is positive. Generally, the logic on which they are based is vague and overly complex for most students. One example is the idea of rewinding and fast-forwarding a videotape. It is probably better to use patterning to show students why the rule makes sense, and then they can use the rule confidently.

PRINCIPLE 4 The distributive property applies to multiplication and addition of integers; that is, $a(b + c) = ab + ac$.

Students are likely simply to assume that this property they learned with whole numbers still holds for integers, but it is important to show that it works using several examples. For example,

$$2 \times (-37) = 2 \times (-30) + 2 \times (-7) \qquad -3 \times (-27) = (-3) \times (-25) + (-3) \times (-2)$$
$$= -60 + (-14) \qquad\qquad\qquad\qquad = 75 + 6$$
$$= -74 \qquad\qquad\qquad\qquad\qquad\qquad = 81$$

Dividing Integers

Principles for Dividing Integers

As with multiplication of integers, although the principles for dividing integers are fairly easy for students to learn, explaining why those principles make sense is a greater challenge.

PRINCIPLES FOR DIVIDING INTEGERS

1. Division of integers can be modelled using a sharing, a grouping (measurement), or an inverse multiplication meaning.
2. The rules for assigning signs to the quotient of two integers are based on the rules for products.
   - The quotient of two positives or two negatives is positive.
   - The quotient of a positive and a negative is negative.

PRINCIPLE 1 Division of integers can be modelled using a sharing, a grouping (measurement), or an inverse multiplication meaning.

One of the meanings of a quotient like $10 \div 2$ is the size of the share if 2 people share 10 items. That approach can also be used to describe the quotient of a negative divided by a positive. For example,

Dividing $(-10) \div (+2)$

MODELLING THE SHARING MEANING WITH COUNTERS

If 10 blue counters (-10) are shared by 2 people $(+2)$, each gets 5 blue counters (-5).

$$(-10) \div (+2) = (-5)$$

MODELLING THE SHARING MEANING ON A NUMBER LINE

If a jump on the number line of 10 to the left (-10) is divided into 2 jumps $(+2)$, each jump is 5 to the left (-5).

$$(-10) \div (+2) = (-5)$$

Another meaning of division is grouping or measurement. It is useful for explaining why a positive divided by a positive, or a negative divided by a negative is positive.

Dividing (−10) ÷ (−2) Using the Grouping Meaning

MODELLING WITH COUNTERS

(−10) ÷ (−2) can mean:

How many groups of 2 blue counters (−2) are in 10 blue counters (−10)?
There are 5 groups, which is (+5).

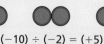

(−10) ÷ (−2) = (+5)

MODELLING ON A NUMBER LINE

(−10) ÷ (−2) can mean:

How many jumps of (−2) are in a jump of (−10)?
There are 5 jumps, which is (+5).

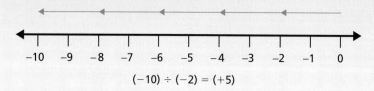

(−10) ÷ (−2) = (+5)

A third meaning of division is inverse multiplication. The quotient of two numbers is the number you must multiply the divisor by to get a product equal to the dividend. For example, if $(-10) \div 2 = \blacksquare$, then $2 \times \blacksquare = (-10)$. It is this meaning of division that best explains the quotient of integers with different signs.

Dividing Using the Inverse Multiplication Meaning

$(-10) \div (+2) = \blacksquare \rightarrow (+2) \times \blacksquare = (-10)$

Using the rules for multiplying, it is clear that the missing number is negative.

Since $(+2) \times (-5) = (-10)$, then $(-10) \div (+2) = (-5)$.

$(-10) \div (-2) = \blacksquare \rightarrow (-2) \times \blacksquare = (-10)$

Using the rules for multiplying, it is clear that the missing number is positive.

Since $(-2) \times (+5) = (-10)$, then $(-10) \div (-2) = (+5)$.

PRINCIPLE 2 The rules for assigning signs to the quotient of two integers are based on the rules for products.

• The quotient of two positives or two negatives is positive.

• The quotient of a positive and a negative is negative.

Because of the relationship between multiplication and division, the sign rules are the same for division as for multiplication.

For example,

- $(-12) \div (-4)$ must be $(+3)$, since $(-4) \times (+3) = (-12)$
- $(-12) \div (+4)$ must be (-3), since $(+4) \times (-3) = (-12)$
- $(+12) \div (-4)$ must be $(+3)$, since $(-4) \times (+3) = (-12)$

Chapter 3

Geometry

Shapes and Their Properties

Geometry is one, but not the only, aspect of mathematics where visualization is important. Whiteley (2004) speaks about visual reasoning as "seeing to think." Because visualization is such an obvious aspect of geometry, using geometric thinking as a tool to improve visual reasoning makes sense.

BIG IDEAS FOR SHAPES AND THEIR PROPERTIES

1. Some attributes of shapes are quantitative, others are qualitative. For example, the fact that hexagons are shapes that will tile is qualitative; the fact that hexagons have six vertices is quantitative.

2. Many of the properties and attributes that apply to 2-D shapes also apply to 3-D shapes.

3. How a shape can be cut up (dissected) and rearranged (combined) into other shapes helps us attend to the properties of the shape, for example, where right angles are and where the equal sides are.

4. Many geometric properties and attributes of shapes are related to measurement, for example, how far apart two parallel lines are or how the side lengths of a shape compare.

A knowledge of big ideas can help teachers choose, shape, and create tasks and use questioning to help students make powerful connections.

Each teaching idea in this section of the chapter will indicate which Big Idea(s) for Shapes and Their Properties (BISP) can be emphasized.

Geometric Attributes and Properties of Shapes

As students become more familiar with geometric attributes, they gradually gain an awareness of the specific attributes that define each class of shape, that is, the properties of shapes. A property is an attribute that applies to all the shapes of a certain class. For example, the class of shapes called rectangles has these properties: four straight sides, two pairs of equal sides, and four right angles. Quadrilaterals (a class of shapes that includes rectangles) also have four straight sides, but only some quadrilaterals have four right angles. So, having four right angles is a property of all rectangles, but not of all quadrilaterals.

A property is an attribute that applies to all the shapes of a certain classification.

Using Properties to Classify 3-D Shapes

Polyhedrons, Spheres, Cones, and Cylinders

Teaching Idea | 3.1

Students can solve riddles like these about 3-D shapes:

- A 3-D shape has eight vertices. What could it be? [quadrilateral-based prism, including a cube or a heptagon-based pyramid]

- A 3-D shape has six more edges than vertices. What could it be? [e.g., hexagon-based pyramid]

Focus on BISP 1, about quantitative properties of shapes, by asking: *Choose a 3-D shape. Create a riddle that uses one or more numbers.* [e.g., a 3-D shape has nine fewer vertices than edges. What could it be? [a decagon-based pyramid]

Components are the individual parts that go together to make a shape.

A polyhedron is any 3-D shape whose faces are all polygons. Polyhedrons can have as few as four faces (a triangle-based pyramid) or they can have many more. Each face can have as few as three sides. All prisms and pyramids are polyhedrons (or polyhedra). Spheres, cones, and cylinders are not polyhedrons since their surfaces are not polygons, but they are shapes students are expected to become familiar with.

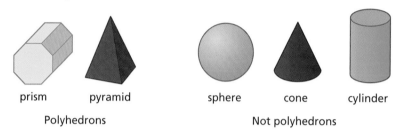

| prism | pyramid | sphere | cone | cylinder |

Polyhedrons   Not polyhedrons

Components of 3-D Shapes

In order to describe the properties of 3-D shapes, and to classify them, students need to understand their components. Components are the individual parts that go together to make a shape—faces, curved surfaces, edges, curved edges, and vertices. Concrete models of 3-D shapes allow students to explore these components in a very hands-on way. For example, students would be able to manipulate a model of a triangle-based prism to view, touch, and count all five faces—two triangles and three rectangles. They would also be able to touch and count each of the six vertices and nine edges.

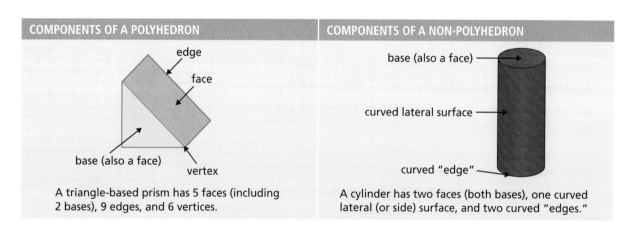

COMPONENTS OF A POLYHEDRON	COMPONENTS OF A NON-POLYHEDRON
edge, face, base (also a face), vertex	base (also a face), curved lateral surface, curved "edge"
A triangle-based prism has 5 faces (including 2 bases), 9 edges, and 6 vertices.	A cylinder has two faces (both bases), one curved lateral (or side) surface, and two curved "edges."

One way to familiarize students with 3-D shapes is to provide information about the properties of a particular shape (or have students provide this information for one another), and have students figure out what the shape might be. Some students might be able to do this by visualizing; others will need concrete materials to help them.

Activities with 3-D shapes can often provide realistic and meaningful reasons for studying 2-D shapes. In fact, many teachers prefer to begin their study of shapes with 3-D shapes, which provides a natural context for studying the 2-D shapes that form their faces or surfaces.

Using Properties to Classify 2-D Shapes

Polygons and Circles

A polygon is a closed 2-D shape whose sides are straight line segments that intersect only at the vertices. A circle is a set of points (the circumference) that are the same distance from a given point (the centre).

Components of 2-D Shapes

To describe the properties of 2-D shapes, and to classify them, students can describe their components.

Teaching Idea | **3.2**

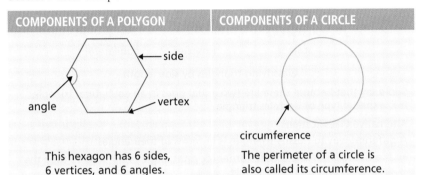

COMPONENTS OF A POLYGON	COMPONENTS OF A CIRCLE
This hexagon has 6 sides, 6 vertices, and 6 angles.	The perimeter of a circle is also called its circumference.

Ask students to create a 2-D shape collage to fit a set of specific rules, such as

- All the shapes are quadrilaterals.
- Two shapes have parallel sides, but are not rectangles.
- Three shapes have lines of symmetry.

Focus students on BISP 1 by asking: *Choose a different property of shapes that does not involve using numbers. Describe the shapes in your collage in terms of that property.* [e.g., three of the shapes have some equal sides]

Classifying Polygons

NAMING POLYGONS BY NUMBER OF SIDES

Sample Polygons	Number of Sides	Name	Sample Polygons	Number of Sides	Name
	3	triangle		8	octagon
	4	quadrilateral		9	nonagon
	5	pentagon		10	decagon
	6	hexagon		11	hendecagon
	7	heptagon		12	dodecagon

Note that, although most polygons are called "__gons," triangles and quadrilaterals are not. That makes it less likely for some students to consider these shapes as types of polygons.

An interesting property of polygons is that the number of vertices is always equal to the number of sides.

Classifying Regular Polygons

REGULAR POLYGONS

In a regular polygon, all the sides are the same length and all the angles are the same size. Equilateral triangles and squares are examples of regular polygons.

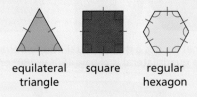

equilateral triangle · square · regular hexagon

PROPERTIES OF REGULAR POLYGONS

- All regular polygons are convex shapes that have all sides equal in length and all angles equal in size.
- As the number of sides on a regular polygon increases, so does the angle at each vertex. The result is a series of polygons that look more and more like a circle.

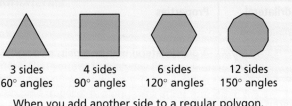

| 3 sides | 4 sides | 6 sides | 12 sides |
| 60° angles | 90° angles | 120° angles | 150° angles |

When you add another side to a regular polygon, each angle increases by the same amount.

Planes, Lines, and Angles

To describe geometric situations, students often need to understand what planes, lines, line segments, rays, and angles are.

Planes

Although planes are often not addressed in K to 8 formally, students discuss planes informally when working with many aspects of geometry. A plane is a 2-D or flat surface that goes on forever in two directions that are not on the same line. The faces of 3-D shapes are parts of planes. Planes can be parallel or they can intersect. The term "plane" may come up in talking about planes of symmetry (see **page 104**).

RELATING PLANES AND 3-D SHAPES	EXAMPLE
• Each face is on a different plane. • Edges are where two planes meet or intersect. • Vertices are where three or more planes meet or intersect.	Adjacent faces of a cube are parts of intersecting perpendicular planes, and opposite faces are parts of parallel planes. 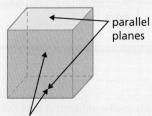 parallel planes intersecting planes Cubes consist of faces that are on parallel planes and on intersecting perpendicular planes.

Lines, Segments, and Rays

It is normally taken for granted that students know what lines are, but the term is often used interchangeably with ray and line segment. The usual way of naming a line, line segment, or ray is to locate and name two points on it (each usually indicated with a dot). The standard convention for showing the difference between lines, line segments, and rays is to use an arrow to indicate where a line extends indefinitely and to use a dot to indicate an endpoint.

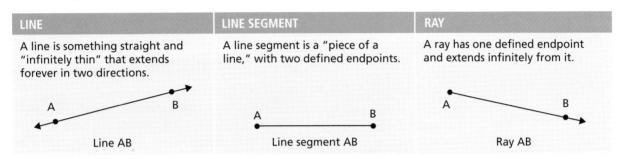

LINE	LINE SEGMENT	RAY
A line is something straight and "infinitely thin" that extends forever in two directions.	A line segment is a "piece of a line," with two defined endpoints.	A ray has one defined endpoint and extends infinitely from it.

Line AB

Line segment AB

Ray AB

Lines, line segments, and rays can be parallel or they can intersect. Parallel lines, line segments, and rays never meet, since they remain a constant distance apart. Whenever two lines intersect, they meet at a single point. When lines intersect at a 90° angle, they are said to be perpendicular.

PARALLEL LINES	INTERSECTING LINES	PERPENDICULAR LINES
Parallel lines do not meet. The distance between parallel lines remains constant.	Intersecting lines meet at a single point. The distance between the lines increases as you move away from that point.	Perpendicular lines are intersecting lines that meet or cross at a right angle.

Symmetry

There are two types of symmetry in 2-D and 3-D geometry—reflective (or line or reflectional or mirror) symmetry and rotational symmetry (or turn symmetry).

- When 2-D shapes are divided along one or more lines of symmetry, or 3-D shapes are divided across one or more planes of symmetry, and the opposite sides are mirror images, you say that the shapes have reflective symmetry.
- Rotational symmetry refers to the number of times a 2-D shape fits over an image of itself when it is rotated a full rotation, or the number of times a 3-D shape appears exactly the same during a full rotation.

Reflective Symmetry of 2-D and 3-D Shapes

A shape has reflective symmetry if one half of the shape is a reflection of the other half. Both 2-D and 3-D shapes can have reflective symmetry. In a 2-D shape, the reflection occurs across a line. In a 3-D shape, it occurs across a plane.

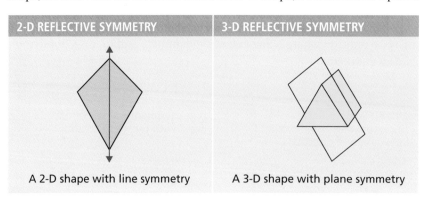

2-D REFLECTIVE SYMMETRY	3-D REFLECTIVE SYMMETRY
A 2-D shape with line symmetry	A 3-D shape with plane symmetry

Line Symmetry

When one half of a shape reflects onto the other half across a line, the line is called the line of symmetry. Shapes can have one or more lines of symmetry. In fact, a circle has an infinite number of lines of symmetry.

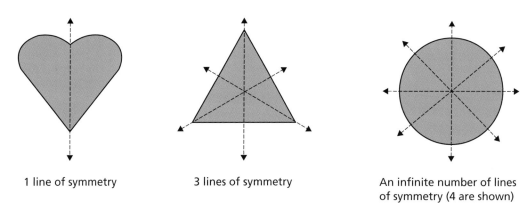

1 line of symmetry   3 lines of symmetry   An infinite number of lines of symmetry (4 are shown)

Teaching Idea | 3.7

To focus on BISP 1, ask: Tell me something that is true about this shape that does not involve using numbers. [e.g., it's symmetrical] Tell me something that is true about the shape that does involve a number. [e.g., it has eight lines of symmetry]

Students are often surprised to find that a shape can have more than one line of symmetry. Note that the more sides there are on a regular polygon, the more lines of symmetry there will be. This is because the shape is getting more and more like a circle. In addition, the number of lines of symmetry in a regular polygon is always equal to the number of vertices.

Students can test the symmetry of a 2-D shape either by folding it to see if the halves match, or by placing a transparent mirror (or a commercial tool called a Mira) on the shape. In each case, if the shape is symmetrical along the fold line or where the Mira has been placed, the image of one side of the shape will fall right on top of the other side of the shape.

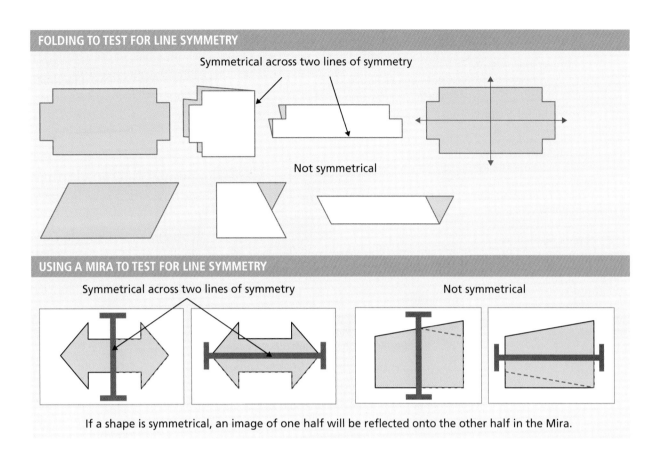

FOLDING TO TEST FOR LINE SYMMETRY

Symmetrical across two lines of symmetry

Not symmetrical

USING A MIRA TO TEST FOR LINE SYMMETRY

Symmetrical across two lines of symmetry

Not symmetrical

If a shape is symmetrical, an image of one half will be reflected onto the other half in the Mira.

Constructing Shapes and Designs with Line Symmetry

There are a number of different ways for students to construct shapes and designs with line symmetry. Students can use square tiles or pattern blocks, folded paper, a Mira, a geoboard or grid paper, or technology tools such as a drawing program or dynamic geometry software.

Some people use the term "construct" for geometric drawings that use only a straight edge and a compass (and sometimes a Mira or geometry software). The term "construct" is used in this section of the chapter to simply mean create or make.

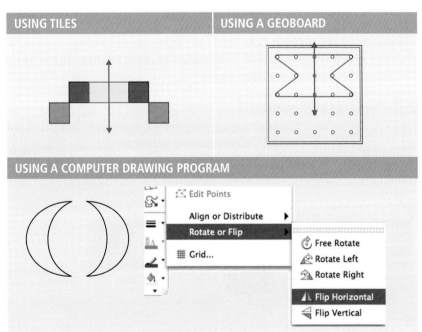

USING TILES

USING A GEOBOARD

USING A COMPUTER DRAWING PROGRAM

Edit Points

Align or Distribute

Rotate or Flip

Grid...

Free Rotate

Rotate Left

Rotate Right

Flip Horizontal

Flip Vertical

Plane Symmetry

When one half of a 3-D shape reflects onto the other half across a plane, the plane separating the halves is called the plane of symmetry. This is why reflective symmetry in 3-D shapes is called plane symmetry.

Teaching Idea | **3.8**

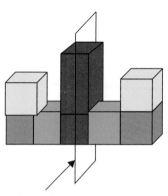

A plane of symmetry

Have students work in pairs. Provide one student with half of a 2-D shape and the other student with half of a cube structure. Tell them to complete their shape so that the other half is the mirror image of what they have.

To focus on BISP 2, ask: *How is what you did to complete your shape like what your partner did?* [Both of us made something on the other side to match what was on the first side; if a part of the first side was close to the line or plane, so was its matching part of the other side.]

Note that this cube structure has a second plane of symmetry that is perpendicular to the one shown above. It bisects all of the cubes vertically.

It is sometimes difficult to test 3-D shapes for symmetry, since the shapes cannot be folded and overlapped. However, they can often be represented with modelling clay or linking cubes, and then divided in half. Then students can examine and manipulate the halves to see if one half looks like the exact "reverse" of the other half.

There are a number of different ways for students to construct shapes with plane symmetry. These include building structures with linking cubes, stacking shape blocks, and using modelling clay. Pattern block stacks can be tested for symmetry with a Mira, while modelling clay shapes can be cut through the centre with a strand of dental floss.

Rotational Symmetry of 2-D and 3-D Shapes

A shape has rotational symmetry if, when you turn it around its centre point (2-D) or an axis of rotation (3-D), it fits into an outline of itself at least once before it has completed a full rotation.

2-D Rotational Symmetry

One way to test for rotational symmetry is to trace the shape, and then turn the tracing over the original shape around a pencil point to see whether it fits over itself. For example, a rectangle fits over itself twice—once after a half turn, and again after a complete turn.

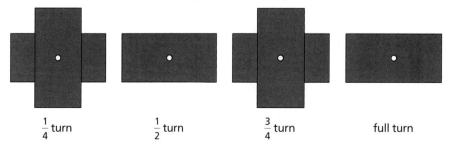

$\frac{1}{4}$ turn $\frac{1}{2}$ turn $\frac{3}{4}$ turn   full turn

A rectangle has rotational symmetry because its image fits more than once in a full turn.

Line Symmetry Versus Rotational Symmetry

Rotational symmetry is distinct from reflective symmetry. A parallelogram has rotational symmetry but no reflective symmetry. Similarly, a shape can have reflective symmetry without having rotational symmetry. An example is this isosceles triangle.

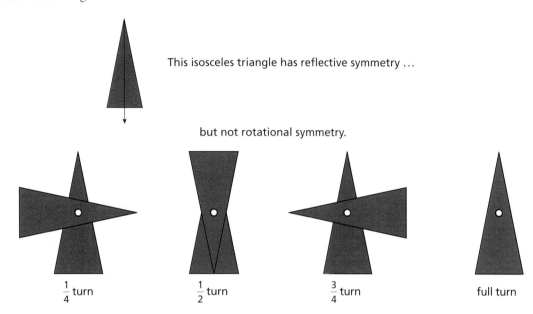

This isosceles triangle has reflective symmetry …

but not rotational symmetry.

$\frac{1}{4}$ turn $\frac{1}{2}$ turn $\frac{3}{4}$ turn   full turn

An isosceles triangle (non-equilateral) does not have rotational symmetry because its image does not fit on itself until it has completed a full turn.

Although the two types of symmetry are distinct, there are relationships. For example, if a shape has two or more lines of symmetry, it also has rotational symmetry.

Measuring 2-D Rotational Symmetry

The number of ways that a shape fits over its outline is called its order of rotational symmetry. For example,

- For an equilateral triangle, the order of rotational symmetry is three because it fits over its image three times within a full turn.
- For a circle, the order of rotational symmetry is infinite, because it fits over its image an infinite number of times as it turns through 360°.
- A shape with no rotational symmetry has an order of rotational symmetry of one because it only fits once in a full rotation (at the end of the full rotation).

3-D Rotational Symmetry

Rotational symmetry is also associated with 3-D shapes. For example, a cube has 13 axes of symmetry—lines around which the shape can be rotated to reproduce its original orientation before it has completed a full turn. In the case of a cube, there are four axes that connect pairs of opposite vertices, six axes that connect the midpoints of opposite edges, and three axes that connect the centres of opposite faces, as shown on the next page.

Teaching Idea | **3.9**

Ask students to think of a real world object that has a flat surface to fit each description:

- rotational symmetry, but no mirror symmetry
- mirror symmetry, but no rotational symmetry
- both rotational and mirror symmetry
- neither rotational nor mirror symmetry

To focus on BISP 1, ask: *How many lines of symmetry does a 2-D shape have if it has both rotational and mirror symmetry?* [two or more]

Three of 13 Axes of Symmetry of a Cube

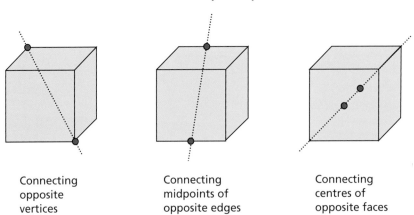

Connecting
opposite
vertices

Connecting
midpoints of
opposite edges

Connecting
centres of
opposite faces

Representing Shapes

Creating representations of shapes is a good way for students to use their visualization skills. It is also an area of geometry that is closely linked to other curriculum areas, especially the visual arts.

Representations can take many forms, including modelling clay or linking cube models, pattern blocks, skeletons, nets, and various types of drawings. As students develop geometrically, their representations will appear more and more like the real shapes they represent. As well, they will be able to create more sophisticated representations, for example, moving from simple sketches of 3-D shapes to isometric drawings.

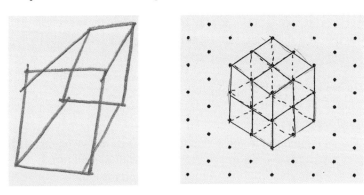

A simple sketch of a 3-D shape and an isometric drawing

Concrete Models

One way to represent shapes is to make concrete models. Although this is obviously the case for primary students, Grades 4 to 8 students also benefit from building models with, for example, pattern blocks or linking cubes.

Skeletons

Another type of model is a skeleton—a physical representation of just the edges and vertices of a 3-D shape, or just the sides and vertices of a 2-D shape. Materials for constructing skeletons include toothpicks and small

---

Teaching Idea | **3.10**

Have students stack pattern blocks to build prisms like the one below.

To focus on BISP 1, ask:
Which of your shapes are symmetrical? [all of them]
Which of your shapes has the most faces? [hexagon-based prism]

Teaching Idea | **3.11**

Have students build rectangle-based prisms with linking cubes like the one below.

To focus on BISP 4, ask:
Arrange your cubes to make other rectangle-based prisms. Why can you make more prisms if your base has 12 cubes than if it is has 11 cubes? [12 has more factors than 11, so there are more ways to make the base.]

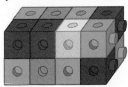

balls of modelling clay, or straws connected with bent segments of pipe cleaners. Toothpicks are especially useful for modelling regular shapes because they have a uniform length. There are also commercial construction toys, such as Tinkertoy, K'nex, Zoob, Geomag, and D-stix, that are suitable for constructing skeletons.

A skeleton of a cube

A skeleton of a regular hexagon

In order to construct the skeleton of a shape, many students need to have the shape in front of them. This way, they can look at and touch the edges and vertices (focusing on the properties of the shape), to develop a mental picture of how many there are and where they belong. Others are comfortable working from a picture of the shape, and still others can sometimes work directly from verbal descriptions.

Skeletons help students see familiar shapes in a different way. When they are working with solid shapes, students often tend to focus more on the faces than on other components. The process of making a skeleton, where the faces are implicit, helps students become more aware of other components: sides, edges, vertices, and angles. It also helps them create a mental image of the shape, which will stay with them even when they no longer have concrete models to look at. For example, when asked for the number of edges on a cube, a student might visualize the cube skeleton and mentally count the edges.

Teaching Idea 3.12 shows how a task involving skeletons can be used to help students learn more about the edge and vertex properties of 3-D shapes. For this activity, you may want to provide some samples of 3-D shapes for students' reference.

Nets

A net is a 2-D representation of a 3-D shape that can be folded to re-create the shape. When students make nets, they focus particularly on the faces, and how the faces fit together to form the shape. It is important for students to realize that there are often many different nets for a single shape. Even though the faces do not change, they can be connected in different ways. For example, all of the nets below can be folded to make a cube.

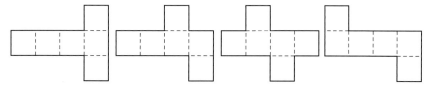

Possible nets for a cube

Teaching Idea | **3.12**

Provide students with ten long straws, ten short ones, and pipe cleaner connectors. Ask students to make three different prisms and pyramids using the materials. Then ask them to name some shapes that are impossible [e.g., hexagon-based prism].

To focus on BISP 1, ask: *How many different pyramids could you make if you use all straws of the same length? Why?* [only a triangle-based pyramid; if I try to make a base with more edges, the straws that go up to the point are not long enough]

Creating Nets

Students might roll a shape on a piece of paper and trace each face to create a net or they might consider properties of the shape to create the net. In creating the net for a cube, students work with the following properties of a cube:

- 6 congruent square faces
- 3 pairs of opposite parallel faces
- 3 faces joining at each vertex, and
- congruent edges that meet at right angles

However, students cannot assume that because a cube has six square faces, any arrangement of six squares will create a net. The patterns below are made from six squares, but they are not nets for a cube because there is no way to fold them to create a top face, a bottom face, and four lateral faces.

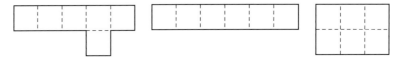

Not nets for a cube

For pyramids, they might draw the base and then draw congruent isosceles triangles on each side, making sure that the height of each triangle is sufficient to reach farther than the centre of the base. Many students are surprised to find that pyramids with different heights can be created on the same base.

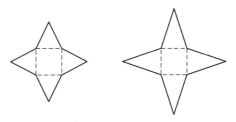

Nets for two different pyramids with the same base

Teaching Idea | **3.13**

Provide each student with an empty paper towel roll or toilet paper roll. Tell them to imagine a cylinder of the same shape and size (but with closed bases). Ask students to create a net for the cylinder and then assemble it to test it.

To focus on BISP 4, ask: *What measurements on your net did you have to make to be sure it worked?* [diameters of the circles and length of the rectangle had to match distance around each circle]

Nets are sometimes more difficult to make when the shape being modelled is not a polyhedron, that is, when some of the edges are curved. In the net for a cylinder, it is important for students to ensure that the width of the rectangle representing the lateral curved surface matches the circumference of the circular base.

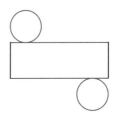

For a cylinder net to work, the red parts must be the same length.

Drawing 3-D Shapes

Another way to represent a shape is with a picture. Before students can draw a shape, they need time to examine it. For example, to make perspective drawings of prisms, they need to focus on parallel bases and edges, as well as on face shapes.

Isometric Drawings

An isometric drawing does not distort parallel lines the way a perspective drawing does. It is drawn on isometric dot paper, which is sometimes called triangle dot paper. Edges that are parallel on the original shape are represented by parallel lines in the drawing. Isometric drawings of cubes and linking cube structures are the easiest for students to draw first. However, with experience, they can learn to create isometric drawings of more complex 3-D shapes.

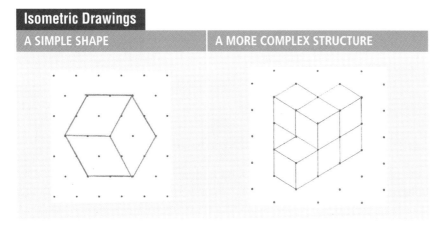

Isometric Drawings

| A SIMPLE SHAPE | A MORE COMPLEX STRUCTURE |

Orthographic Drawings

Orthographic drawings show views of a shape from the front, top, and sides. Drafters often create various views of a 3-D shape in order to help someone else build it.

When students are ready to begin drawing views of their own, they might start with a simple 3-D shape, such as a prism. View drawings can also be done for shapes with curved surfaces. When surfaces are curved, the view shows a flat representation of what the curved surface looks like.

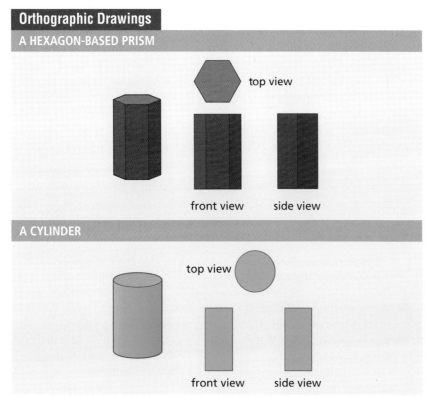

Orthographic Drawings

A HEXAGON-BASED PRISM

top view

front view   side view

A CYLINDER

top view

front view   side view

Base Plans

A base plan is another way of mapping a cube structure. The base plan is a bird's-eye view of the structure's base that uses numbers to indicate the height of each part of the structure. For example, the base plan below shows that the two rows at the back of the structure are three cubes high, while the front row is only one cube high.

3	3
3	3
1	1

front

Combining and Dissecting Shapes

Teaching Idea | **3.15**

To focus on BISP 3, present the following challenges:

- combine two isosceles triangles to create a square

- combine two different quadrilaterals to create a triangle

- combine two congruent hexagons to create an octagon

Then ask questions such as *What do you notice about the triangles you combined to make a square? Would another type of triangle have worked?* [congruent right triangles; no, each triangle needs a right angle and two 45° angles to make the square]

One of the key concepts in geometry is that any shape can be created by either combining or dissecting other shapes. Students are working with this concept when they are putting shapes together to make another shape, a picture, a design, or a structure, or when they are cutting a shape into pieces.

As students combine and dissect shapes, they are learning about properties of shapes. They can learn, for example, that diagonals cut a rectangle into four parts that have equal areas that are not all congruent unless the rectangle is a square.

The diagonals of a rectangle dissect it into two pairs of congruent triangles.

The diagonals of a square dissect it into four congruent triangles.

The ability to combine and dissect shapes also supports the development of measurement formulas. For example, cutting a parallelogram apart and reassembling it to form a rectangle can help students see why the area of a parallelogram can be determined by multiplying its base length by its height.

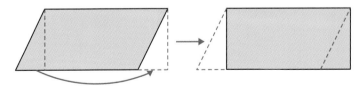

Dissecting a parallelogram to show why its area is equal to the area of a rectangle with the same base/length and height/width.

Students might be offered challenges that involve combining different types of shapes to create other shapes. The combinations should require students to consider properties of the shapes they are combining as well as the resulting shape.

Using Tangrams

Tangram pieces (sometimes called "tans") are formed by dissecting a square into seven shapes as shown in **Teaching Idea 3.16**. The pieces can then be combined to reconstruct the original square, as well as to create many other shapes.

Tangrams can be used to illustrate shape combinations and shape dissections. At the Grade 4 to 8 level, puzzles that are not the actual size of the tangrams (either enlarged or reduced from the real size), might be most appropriate.

Dissecting Shapes

The idea that shapes can be dissected, or divided into parts, is fundamental to many geometry concepts that students will explore in the intermediate and senior grades, as shown in the examples that follow.

SAMPLE TASKS INVOLVING DISSECTING SHAPES

What shapes can you make by cutting a square into four pieces?

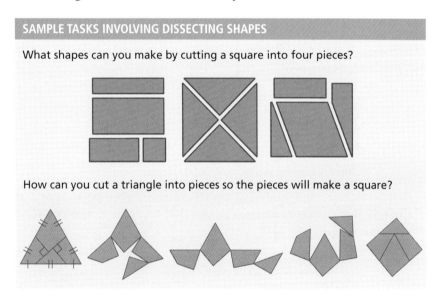

How can you cut a triangle into pieces so the pieces will make a square?

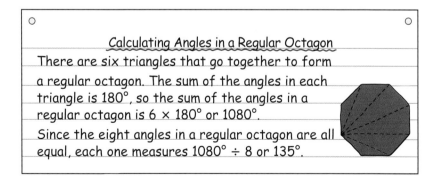

Dissecting to Calculate Angles

Students will learn that they can calculate the angle size for any regular polygon by dissecting the shape into triangles.

> ### Calculating Angles in a Regular Octagon
> There are six triangles that go together to form a regular octagon. The sum of the angles in each triangle is 180°, so the sum of the angles in a regular octagon is 6 × 180° or 1080°.
> Since the eight angles in a regular octagon are all equal, each one measures 1080° ÷ 8 or 135°.

Teaching Idea | **3.16**

To focus on BISP 3, present students with the following challenge:

Here are the seven tangram shapes arranged in a square.

Can you make a square with six tans? [no] Can you make a square with two, three, four, five, or seven tans? [yes]

Then ask: *What do you notice about the tans that work?* [When I put them together, I can make 90° angles; 180° angles, and 360° angles.]

Teaching Idea | **3.17**

To focus students on BISP 3, have them make a triangle congruent to the largest tan using any or all of the remaining six pieces.

Then ask: *What do you notice about the shapes that work?* [When I put them together, I can make 90° angles and 180° angles.]

Congruence and Similarity

Teaching Idea | **3.18**

An interesting way to explore congruence is through optical illusions.

Focus on BISP 4, by asking: *Which of the inner circles below is larger? Which of the vertical line segments is longer?* [They are the same.]

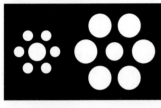

Congruence and similarity are geometric concepts that apply to both 2-D and 3-D shapes, although work with these concepts at the K to 8 level is mainly limited to 2-D shapes.

Two shapes are considered to be congruent if one can be transformed into the other through a series of flips, slides, and/or turns.

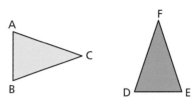

Congruent triangles

The word "congruent" is not only used to describe whole shapes, but also specific components of shapes, such as sides and angles. For example, these two triangles have a pair of congruent sides and a pair of congruent angles, but they are not congruent shapes.

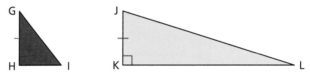

These triangles have a pair of congruent sides and a pair of congruent angles.

Congruence can also be used to describe the properties of a single shape. For example,

An equilateral triangle has three congruent sides and three congruent angles.

Two shapes are said to be similar if they have the same shape, and the corresponding sides are in proportion to one another. Congruent shapes are similar, but so are two shapes where one is an enlargement or reduction of the other.

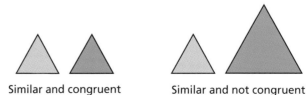

Similar and congruent   Similar and not congruent

Congruence of 2-D Shapes

Two 2-D shapes are congruent if they are identical in shape and size—that is, if one is an exact duplicate of the other. Students sometimes do not understand the difference between the math term "congruent" and the everyday term. It is important to recognize that, in mathematics, "congruent" applies only to shape and size. Thus, shapes can be different colours, or oriented in different ways, and they will still be congruent as long as they are the same shape and the same size.

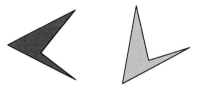

These shapes may be congruent but they are not exactly the same—the colour and orientation are different.

Determining Congruence of 2-D Shapes

There are many ways to test for congruence, some more sophisticated than others.

SUPERIMPOSING	MEASURING CORRESPONDING SIDES AND ANGLES
The easiest way to test for congruence is to put one shape on top of the other to see if they match.	A more sophisticated strategy (and one that works for shapes in any orientation) is to measure the sides and angles of both shapes and compare the results. On congruent shapes, all the corresponding side lengths and angle measures will match.

Constructing Congruent 2-D Shapes

Once students understand the nature of congruence, they can begin making congruent shapes of their own, using a wide variety of materials.

USING A GEOBOARD	MEASURING
Students can use geobands to form congruent shapes on a geoboard. Using a geoboard helps to focus students on the fact that congruent shapes have the same side lengths.	Once students understand that congruent shapes have congruent side lengths and angle measures, they can use rulers and protractors to construct congruent shapes.

Teaching Idea | **3.19**

Eventually we want students to know how to provide the right amount of information for someone to construct a shape congruent to a given one, for example:

• For a square, all you need to know is the fact that it is a square and one side length (two pieces of information).
• For a parallelogram, all you need to know is the fact that it is a parallelogram, two pieces of information about the side lengths (or the base and height), and one angle (four pieces of information).

To focus on BISP 1, ask: *How much information do you need to provide for someone to draw a congruent triangle?* [three pieces of information: three side lengths, or two side lengths and a contained angle]

Students might also use dynamic geometry software to construct shapes that are congruent to given ones.

Congruence of 3-D Shapes

In the case of 3-D shapes, it is much more difficult to determine congruence. If two shapes simply look the same, this may be enough for some students to assume congruence. However, testing congruence requires the ability to use measurements to compare shapes—a characteristic of more mathematically sophisticated students. Structures made of standard linking cubes might be easier to test, since the cubes can be counted in order to make comparisons.

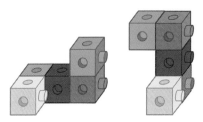

Linking cube structures are good for working with 3-D congruence.

Similarity of 2-D Shapes

Figures that have the same shape, but not necessarily the same size, are said to be similar. Congruent shapes are similar, but so are reductions and enlargements of a shape. When shapes are similar, corresponding angles are congruent, and corresponding side lengths are all enlarged or reduced by the same factor.

In the case of the similar rectangles below, the ratio of the length to the width is the same in both rectangles. This is not true in the non-similar rectangles, where the ratio of length to width for the rectangle on the right is less than for the rectangle on the left.

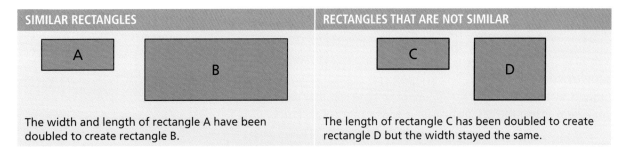

SIMILAR RECTANGLES	RECTANGLES THAT ARE NOT SIMILAR
A B	C D
The width and length of rectangle A have been doubled to create rectangle B.	The length of rectangle C has been doubled to create rectangle D but the width stayed the same.

Teaching Idea | **3.20**

To focus on BISP 4, ask: *One triangle has two 40° angles. Another triangle has a 100° angle and a 40° angle. Could the triangles be similar?* [yes] *Explain your thinking.* [Both triangles have angles 40°, 40°, and 100°.]

Determining Similarity of 2-D Shapes

If two shapes are similar, the following statements are true:

- Corresponding angles in both shapes are congruent.
- You can multiply or divide all the side lengths in one shape by the same amount to describe the corresponding side lengths of the other shape.
- The ratio between one side length and another on one shape is the same as the ratio between the two corresponding side lengths on the other shape.

Angle measures alone cannot be used to test two shapes for similarity. For example, a square and a long, thin rectangle both have four 90° angles, but they are not similar shapes. To determine similarity, students need to look at the side lengths of the shapes. Some tests for similarity are outlined next.

Ways to Determine Similarity

MEASURING SIDE LENGTHS

To test the two rectangles below for similarity, students can measure the side lengths to see if they have been increased or decreased by the same proportion.

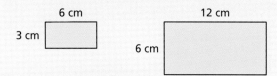

The width and length of the second rectangle are twice as long as the width and length of the first rectangle, so the rectangles are similar.

Another test for similarity involves comparing side lengths within each shape. When you divide the length of each rectangle above by its width, the ratio is 2. Both rectangles are twice as long as they are wide, so the rectangles are similar.

USING DIAGONALS TO TEST RECTANGLES

There is a special and easy way to test rectangles for similarity. First, place the rectangles one on top of the other so the smaller one fits into the bottom left corner of the larger one. Then draw the diagonal of the larger rectangle so it passes through both shapes. If the diagonal of the large rectangle is also a diagonal of the small one, then the rectangles are similar.

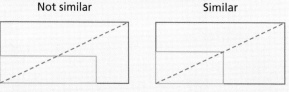

This test works because similar rectangles have the same length to width ratio.

Constructing Similar 2-D Shapes

Once students understand the relationships that make two shapes similar, some may want to try constructing pairs of similar shapes on their own. The methods they use will depend on their mathematical knowledge.

Ways to Construct Similar Shapes

USING SQUARE GRIDS

Square grids are good tools for students to use when they are first learning to construct similar shapes. They can begin by constructing a shape on a square grid, and then copy the shape onto another grid with larger (or smaller) squares.

USING RULERS AND PROTRACTORS

Students may use rulers and protractors to construct similar shapes. For example, to construct a parallelogram similar to a given one, they might measure one angle and examine the ratio of adjacent side lengths, and then copy those measurements to create another parallelogram.

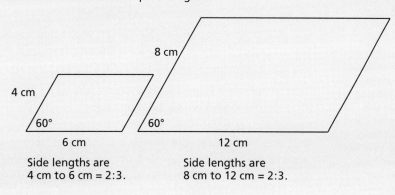

Side lengths are
4 cm to 6 cm = 2:3.

Side lengths are
8 cm to 12 cm = 2:3.

(continued)

Technology makes it possible for students of all ages to construct similar shapes. For example, students could use dynamic geometry software to construct a shape and then reduce it.

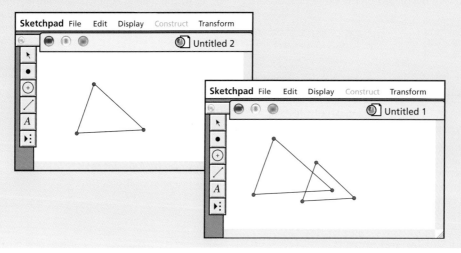

Teaching Idea | **3.21**

Have students draw a right triangle and place non-square rectangles on the three sides. The long side of each rectangle should match the side length of the right triangle and the short side of the rectangle should be half as long. If students calculate each area, they will see that the total area of the rectangles on the legs is the same as the area of the rectangle on the hypotenuse.

Focus on BISP 4 by asking: *Why was it important to measure to create the similar shapes on the sides of the right triangle?* [You need to measure to make sure the ratio of the side lengths is the same.] *Is it always necessary to measure to decide whether two shapes are similar?* [yes, unless you know they are circles or regular polygons]

The Pythagorean Theorem

You can extend the Pythagorean theorem to show an interesting application of similarity. The Pythagorean theorem says that, if the three side lengths of a right triangle are a, b, and c, where a and b are the shorter sides (the legs) and c is the longest side (the hypotenuse), then $a^2 + b^2 = c^2$.

One way to prove the theorem is to place a square on each side (as shown below). You will discover that the total area of the squares on each leg of the right triangle ($a^2 + b^2$) is the same as the area of the square on its hypotenuse (c^2).

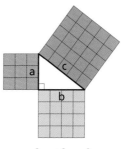

$$a^2 + b^2 = c^2$$

Notice that the three squares on the sides are similar, since their side lengths are proportional: the purple sides are 5 : 5 or 1, the blue sides are 3 : 3 or 1, and the orange sides are 4 : 4 or 1 (in fact, all squares are similar because the ratio of the side lengths of any square is always 1). If you were to place any three similar shapes on the sides of a right triangle, the total area of the two smaller shapes is the same as the area of the larger shape. For example, the total area of the blue and orange half-circles at the top of the next page is equal to the area of the purple half-circle, since the three half circles are all similar. The same is true for the blue, orange, and purple isosceles triangles.

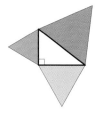

Area of blue shape + Area of orange shape = Area of purple shape

Constructions

Usually in Grade 7 or higher, students are asked to perform geometric constructions. The purpose of the constructions is to allow students to apply or develop an understanding of the properties of shapes. Traditionally, construction tools were a straightedge (a ruler with no numbers on it) and compass, but now constructions are also performed with other tools, such as Miras or dynamic geometry software.

Typically, students might learn to construct

- the perpendicular to a line
- a line parallel to another line
- the perpendicular bisector of a line segment
- a circle with a given centre
- a circle that goes through two or three given points
- an angle bisector

By combining these ideas, students might also learn to construct angles of particular sizes, for example,

- a 45° or 60° angle;
- an equilateral triangle;
- a regular hexagon;
- a circle touching the three sides of a triangle; and so on.

For example, below are shown constructions of a perpendicular to a line using a straightedge and compass, a Mira, and Geometer's Sketchpad.

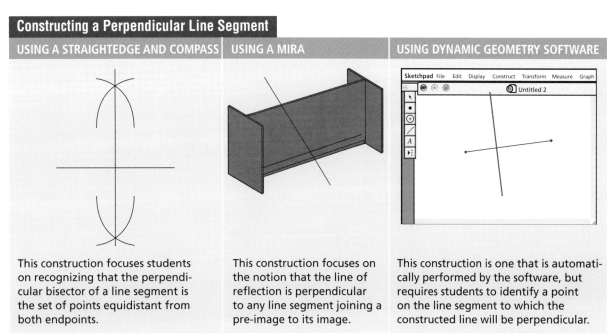

Constructing a Perpendicular Line Segment

USING A STRAIGHTEDGE AND COMPASS	USING A MIRA	USING DYNAMIC GEOMETRY SOFTWARE
This construction focuses students on recognizing that the perpendicular bisector of a line segment is the set of points equidistant from both endpoints.	This construction focuses on the notion that the line of reflection is perpendicular to any line segment joining a pre-image to its image.	This construction is one that is automatically performed by the software, but requires students to identify a point on the line segment to which the constructed line will be perpendicular.

Location and Movement

Geometric experiences involving location and movement support the development of spatial sense and positional vocabulary as students learn to

- describe the positions of objects in structures and pictures
- read and draw maps
- plot points and describe paths on grids
- transform and construct shapes

BIG IDEAS FOR LOCATION AND MOVEMENT

1. Locations can be described using maps, using length and angle measurements, and using coordinate grids.

2. There are three motions, or transformations, that change the position of a shape or possibly its orientation, but do not change its size and shape (translations, reflections, and rotations). Other transformations (like dilatations) can affect size.

3. Transformations have different effects on the position of a shape, so it is often, but not always, possible to look at the original shape and its image to determine which transformation(s) were performed.

4. Transformations are frequently observable in our everyday world. One example context is a tessellation. Transformations are also useful in describing mathematical ideas.

Maps and Coordinate Grids

Maps make it possible to record and describe how objects are located relative to one another. As students develop better spatial sense, their maps better reflect the geometric features of objects in their surroundings and give a more accurate impression of the proportional distances between objects.

Maps might be used to show "static" images of the relative position of objects or might be used to show "active" images or paths. For example, the map below shows how a student would have to move to get from his desk to the door of the classroom. The student might describe his map by saying, "I went forward 5 steps. I turned 90° to the right and went forward 10 steps. I turned 40° to the left and went forward 4 steps. I turned 40° to the right and walked another 10 steps."

Teaching Idea | **3.22**

Show students the grid below and have them find all the paths there are from A to B if you always move right or down. [20 paths] Then have them figure out what all the paths have in common. [all the same length]

Focus on BILM 1 by asking: *How does the grid make it easy to describe a path?* [You can measure the distance of each part of the path by the number of grid units.]

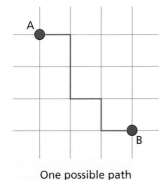

One possible path

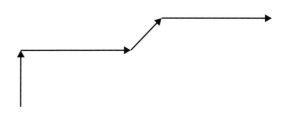

A simple "active" path or map

Working with Grids

At some point, students are ready to use a grid system to identify locations on a map, or to describe how to get from one map location to another.

Places in Alberta

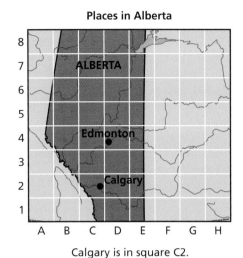

Calgary is in square C2.

Using and interpreting map coordinates is a good introduction to coordinate graphing.

Students can also create designs on grids, and then describe their designs by identifying grid squares to colour. Alternatively, students can be given grid locations and colours and asked to show the design on a grid.

Number Coordinates

After some time working with grids such as city maps, where locations are designated using a letter and a number to describe a horizontal and vertical space in which they are found, students learn to use the conventional number-number code for identifying locations where two grid lines intersect. At this stage, it becomes important for students to recognize that, by convention, the first number indicates the distance from the vertical axis and the second number indicates the distance from the horizontal axis. For example, to reach point $(2, 3)$, you begin at $(0, 0)$, where the two reference axes meet, move two steps to the right, and then three steps up. Designations such as $(0, 0)$ and $(2, 3)$ are called "ordered pairs." Once students are familiar with integers, the coordinate grid is extended to four quadrants. The switch to using labels to describe distance travelled along the axes as opposed to simply labelling the square is not easy for some students.

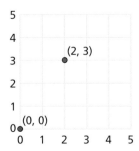

Students should understand that the motivation for the switch from letter–number grids to coordinate grids is to be able to be more specific about location. For example, if I know a location is in B3, I don't know where, within that square, the exact location is. But if I know the location is $(2, 3)$ on the grid, there is no ambiguity about its location.

Provide a grid map labelled with numbers on one axis and letters on the other.

Focus on BILM 1 by asking: *A location on the map is north of G3. Where can it be?* [The row number is 4 or higher. The column letter is probably F, G, or H, or the instructions would have said northeast or northwest.]

Teaching Idea | **3.24**

Ask students to follow this path on a grid and then connect the points in order to identify the shape.

- Plot Point A at (3, 2).
- Go right five to plot Point B.
- Go right two and up two to plot Point C.
- Go up two to plot Point D.
- Go left two and up two to plot Point E.
- Go left five to plot Point F.
- Go left two and down two to plot Point G.
- Go down two to plot Point H.
- Go right two and down two to re-plot Point A.

Students can then create directions for their own secret shape.

Focus on BILM 1 by asking: *How does using coordinates to start guarantee that someone can reproduce your shape?* [Since there is only one point at any ordered pair, they know exactly where to start.]

There are many games and activities to help students learn how to use coordinate systems. Some are commercial games, such as Battleship, while others are teacher-made or student-made. Other games and activities include the following:

- Introduce coordinate graphing by using masking tape to make a large grid on the floor. Use cards to label the axes with coordinates. Invite a student to stand at a point on the grid: "Aaron, please come and stand on point (3, 5)." Then that student can invite another student to stand on a different point, and so on.
- Place objects at different points on the grid and have students use coordinates to describe their locations.
- In a variation of Twister, one student gives a partner directions for placing both feet and both hands on coordinates on a floor grid. Then the students trade roles and play again.

Transformations

Teaching Idea | 3.25

To focus on BILM 4, ask: *How are reflections useful for explaining negative integers?* [Negative integers are reflection images of positive ones if there is a reflection line at zero that is perpendicular to the number line.] *How are reflections useful for explaining why this shape is symmetric?* [One side of the shape is a reflection image of the other side if the reflection line is along the line of symmetry.]

Geometric transformations are motions that affect a shape in some specified way. K to 8 students work with transformations because many mathematical concepts, such as symmetry, are described best using transformations.

Translations, Reflections, and Rotations

There are three transformations that change the location of an object in space, or the direction in which it faces, but not its size or shape. These transformations, called Euclidean transformations, result in images that are congruent to the original object. The three types of Euclidean transformations are translations, reflections, and rotations. In K to Grade 3, students use the more informal but intuitive terms slides, flips, and turns. By Grade 4, students are ready to be introduced to the formal mathematical language of translations, reflections, and rotations.

In the examples that follow, a light green shape with a dashed outline is used to indicate the original shape, a solid green shape is used for the transformation image, and red is used for translation arrows, reflection lines, and turn centres.

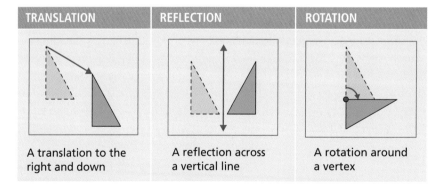

TRANSLATION	REFLECTION	ROTATION
A translation to the right and down	A reflection across a vertical line	A rotation around a vertex

Transformations on Simple Grids

When students first begin learning about translations, reflections, and rotations, they work with concrete shapes on a flat surface. Later, they might work with simple coordinate grids like the one shown at the top of the next page. Working with transformations, especially translations, on a simple grid helps younger students learn to describe motions using mathematical language.

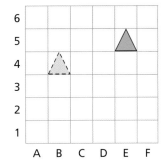

"The green triangle slid 3 spaces right and 1 space up."

Transformations on Coordinate Grids

Later, students begin to explore Euclidean transformations on Cartesian coordinate grids like the one shown below. Here, students can look not only at how a transformation affects the direction in which a shape is facing, but also at how it changes the coordinates of the vertices. For example, the rectangle shown was translated right four units and down one unit. As a result, the first coordinate of each vertex increased by four; the second coordinate decreased by one.

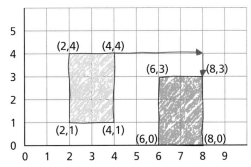

"Each *y*-coordinate decreased by 1 because the rectangle slid 1 unit down. Each *x*-coordinate increased by 4 because the rectangle slid 4 units right."

Translations

A translation moves a shape left, right, up, down, or diagonally without changing the direction in which it faces in any way. This type of transformation is one of the easiest for students to recognize.

Describing Translations

The chart below shows three different translations of the same triangle. Each one is completely defined by a translation arrow that links a point on the original shape to the matching point on the image. Translation arrows could be drawn between each pair of corresponding vertices, but one translation arrow is all that is required to show the translation.

Exploring transformations on simple grids allows students an opportunity to develop and apply positional and movement language.

Teaching Idea | **3.26**

To focus on BILM 4, ask students to look around the classroom for examples of translations. [e.g., when I slide my desk over to work with Andrew, that's a translation; there are reflections in the pattern on my shirt]

Teaching Idea | **3.27**

Tell students that a square with a vertex at (4, 2), another at (4, 6), and a third at (8, 2) was translated to a new position. The image has a vertex at (6, 3).

To focus on BILM 2, ask: *Can you be sure where the image is?* [no] *Could you be sure if you knew that the vertex at (8, 2) moved to (6, 3)? Why?* [yes; since every vertex moves the same: two left and one up.]

A VERTICAL TRANSLATION	A HORIZONTAL TRANSLATION	A DIAGONAL TRANSLATION

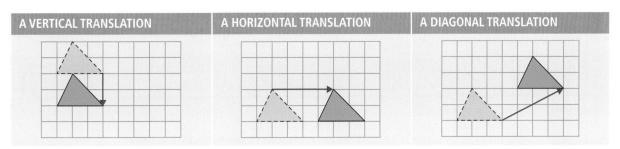

Describing Translation Images

A diagonal translation can be separated into its right–left and up–down components. This is useful when students do translations on coordinate grids.

A common misconception about diagonal translations is that they are performed in two steps, a horizontal motion followed by a vertical motion. Make sure students know that a diagonal slide is one motion in a diagonal direction, even though its description has two parts.

IDENTIFYING TRANSLATION DISTANCE AND DIRECTION

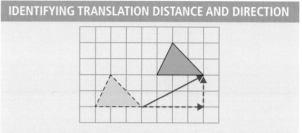

Even though this triangle was translated diagonally, the translation is described as 4 units right and 2 units up.

Mathematicians often use letter notation to show how each vertex on the original shape is matched to a vertex on its translation image.

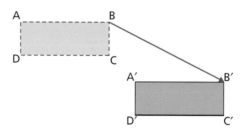

Vertex B′ is the translation image of vertex B.

Properties of Translations

TRANSLATION ARROWS AND PARALLEL SIDE LENGTHS	**ORIENTATION**
• The sides on the translation image are parallel to the corresponding sides on the original shape. • No matter where translation arrows are drawn to link matching points on an original shape and its translation image, these arrows (shown here in red) are all exactly the same length and parallel.	The way a shape faces does not change with a translation. If point C is below and to the right of point A on the original shape, then the translation image of point C also is below and to the right of the translation image of point A on the translation image.
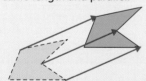	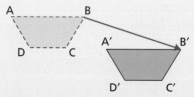
The translation arrows that define a translation are all the same length and are all parallel.	Vertices ABCD and A′B′C′D′ are both clockwise, which shows that a translation image has the same orientation as the original shape.

Reflection

A reflection can be thought of as the result of picking up a shape and flipping it over, as shown below. The reflection image is the mirror image of the original shape.

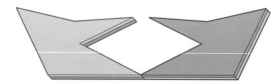

A reflection is like flipping a shape over in space.

Reflection Lines

A reflection is always made across a line called the reflection line. A reflection line can be vertical, horizontal, or diagonal. Reflections are defined by the location and direction of the reflection line.

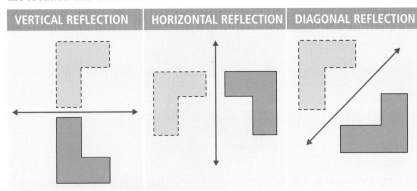

| VERTICAL REFLECTION | HORIZONTAL REFLECTION | DIAGONAL REFLECTION |

Teaching Idea | **3.28**

To focus students on BILM 3, ask: *Can you be sure what single transformation I did to move the shape? Explain.*

[no; it could have been a slide or a flip]

Can you be sure this time? Why?

[yes, it's a flip; the image is facing a different direction]

Properties of Reflections

REFLECTING A SHAPE ACROSS A SIDE

- The only points in the original shape that do not move are those that are located along the reflection line.
- If a 2-D shape is reflected across one of its own sides, the result is a symmetrical 2-D shape.

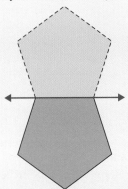

A symmetrical octagon created by reflecting a pentagon across one side

REFLECTING A SHAPE ACROSS A LINE OF SYMMETRY

- If a shape is reflected across one of its own lines of symmetry, then the reflection image will fit exactly over the original shape.

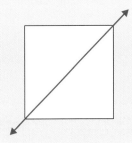

A square reflected across one of its diagonals

DISTANCE FROM THE REFLECTION LINE/ORIENTATION

- Each point on the reflection image is exactly the same distance from the reflection line as its counterpart on the original shape.

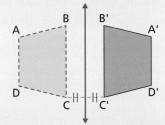

Corresponding points are the same distance from the reflection line.

- Notice that the reflection image has vertices in the the opposite orientation. ABCD are clockwise but A'B'C'D' are counterclockwise.

ANGLE AT THE REFLECTION LINE

- When any point on the original is joined to its counterpart on the reflection image, the connecting line is perpendicular to the reflection line.

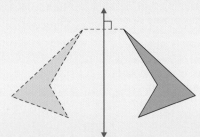

A line connecting any two corresponding points is perpendicular to the reflection line.

Teaching Idea | **3.29**

Students might be interested in knowing that it is always possible to move a shape to a congruent copy in one, two, or three reflections, as shown below. Have them experiment to discover this.

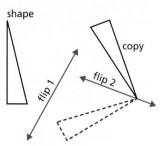

To focus on BILM 2, ask: *Why did the two shapes have to be congruent for this to occur?* [Reflections don't change the size of a shape.]

Teaching Idea | **3.30**

Ask students to reflect a square so its image looks like it might have been a reflection, translation, or rotation, as shown below.

To focus on BILM 2 and 3, ask: *Did the direction the shape was facing change?* [it was supposed to, but it's hard to tell] *How could you check to be sure?* [I could have marked one vertex to see where it ended up.] *Why can you not be sure which transformation was done?* [Since the shape is so symmetric, it is hard to tell which vertex moved where.] *Why do the square's vertices move different distances in a reflection?* [The distance each vertex moves depends on how far it is from the reflection line.]

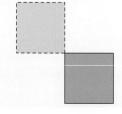

Transparent Mirrors

While students generally have little difficulty reflecting a shape horizontally or vertically, reflections across a diagonal reflection line can be more difficult to perform. In this situation, a transparent mirror (or Mira) can be very helpful. In a transparent mirror, students can actually see the reflection image when they look through the plastic, so it becomes possible to simply trace the reflection image onto a piece of paper.

A transparent mirror is a useful tool for identifying and performing reflections.

A transparent mirror not only allows students to reflect a shape across a diagonal reflection line with ease, it also makes it possible to reflect across a reflection line drawn *through* the original shape. In this case, a student actually has to perform two reflections, first reflecting one side of the shape, and then the other.

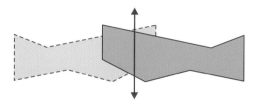

Rotations

A rotation moves a shape in a circle around a turning point, the turn centre. Think of tracing a shape, putting the tracing right on top of the shape, using a pencil tip to hold down the tracing at a particular point, and then rotating the tracing around that point (as shown in the illustration below). When a shape is rotated, only one point, the turn centre, stays fixed; all other points rotate the same way.

Describing Rotations Using Fractions of a Circle

When students first begin working with rotations, they identify them in terms of fractions of a circle: quarter turn, half turn, and three-quarter turn. In addition to describing the amount of rotation, students also need to identify the direction of the rotation (clockwise or counterclockwise). Sometimes clockwise and counterclockwise are abbreviated as "cw" and "ccw."

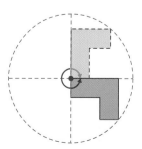

This rotation could be a $\frac{1}{4}$ turn clockwise or a $\frac{3}{4}$ turn counterclockwise.

Describing Rotations Using Angles

Later, students learn that rotations can also be identified in terms of degrees. For example, the figures below show the same shape rotated 90° (a quarter turn) clockwise and 180° (a half turn) counterclockwise. As students rotate shapes, they may find it helpful to think of hands on a clock face. This way, they can focus on the change of position for one side of the shape, as shown below. Although students can generally perform quarter and half turns without a protractor, they will need to measure angles in order to perform other rotations.

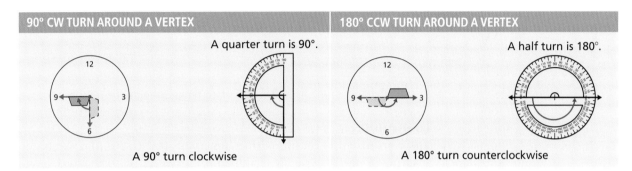

90° CW TURN AROUND A VERTEX	180° CCW TURN AROUND A VERTEX
A quarter turn is 90°.	A half turn is 180°.
A 90° turn clockwise	A 180° turn counterclockwise

Properties of Rotations

CONGRUENT AND SAME ORIENTATION

- Rotation images are congruent to the original image.
- Rotation images have the same orientation as the original shape.

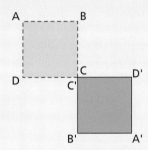

A rotation image has the same orientation as the original shape. As you can see in the example above—vertices ABCD and A'B'C'D' are both clockwise.

EFFECT OF TURN CENTRE

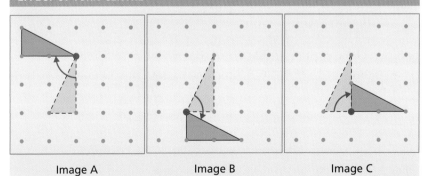

Image A   Image B   Image C

The choice of turn centre determines the location of the rotation image but does not change the direction the triangle is pointing.

Teaching Idea | **3.31**

Ask students to perform two successive reflections in intersecting lines like that shown below.

To focus on BILM 3, ask: *Is there a way to get from the shape to the final image in one motion?* [yes; a rotation with a turn centre at the point where the reflection lines intersect]

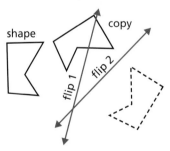

Relating Rotations and Rotational Symmetry

When a shape is rotated around its centre and "lands on" itself at least twice during a full rotation, it has rotational symmetry. For example, a parallelogram (which is not a square) can be rotated around its centre and it fits on its original outline twice during a full rotation.

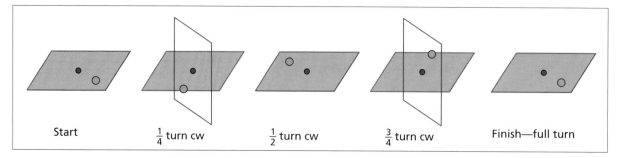

| Start | $\frac{1}{4}$ turn cw | $\frac{1}{2}$ turn cw | $\frac{3}{4}$ turn cw | Finish—full turn |

If you rotate a non-square parallelogram around its centre, it fits two times during a full rotation.

Dilatations

There are also non-Euclidean transformations that change the size of the shape. Only one of these, the dilatation, is introduced, in some jurisdictions, at the K to 8 level. A dilatation increases or decreases the size of the original figure without changing its shape. The result of a dilatation is an image whose angle measures match those in the original, and whose side or curve lengths have all been multiplied or divided by the same amount. This results in an image that is a similar shape. For example, the green triangle below was transformed by doubling the length of each side.

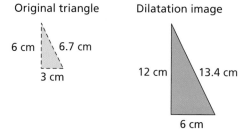

Original triangle   Dilatation image

6 cm   6.7 cm

3 cm

12 cm   13.4 cm

6 cm

In a dilatation, the side lengths all increase or decrease proportionally.

Teaching Idea | **3.32**

Students can explore the properties of dilatations using this method for enlarging a shape:

1. Draw a triangle.
2. Tie two elastic bands together with a knot.
3. Hold down one end near the triangle. Place a pencil at the other end.
4. Use the pencil to move the elastic band so the knot moves along the sides of the triangle. As you move the pencil, you draw a dilatation image of the triangle.

To focus on BILM 2, ask: *Why did the triangle increase in size?* [You doubled the lengths of the sides by doubling up the elastic bands.]

CREATING A DILATATION IMAGE BY MEASURING (SCALE FACTOR 2)

Step 1 Choose a point outside the original shape. This is sometimes called a dilatation point or centre of dilatation.

Step 2 Connect that point to each vertex on the original shape.

Step 3 Measure the length of each connecting line.

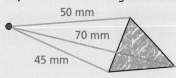

50 mm

70 mm

45 mm

Step 4 Increase or decrease the length of each connecting line using the same proportion (called the scale factor, which in this case is 2).

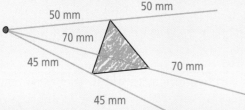

Step 5 Join the ends of the connecting lines to form the new image.

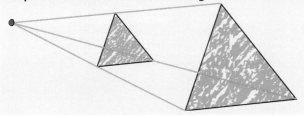

In step 4, the connecting lines were doubled because the scale factor was 2. If the scale factor had been 3, each connecting line would have been tripled resulting in an even larger enlargement. If the scale factor had been 0.5, each connecting line would have been halved resulting in a reduction.

CREATING A DILATATION IMAGE ON A GRID

To create a dilatation image on grid paper, students might count the spaces along each side of the original shape, and then multiply or divide each side length by the same number to determine the lengths of the corresponding sides on the image. In the example on the right, the side lengths of the red rectangle were divided by two to create the green image.

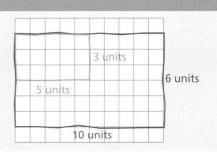

Properties of Dilatations

SIMILARITY

- Dilatations are called similarity transformations because the image is similar to the original shape.

- Every square is a dilatation image of every other square because, by definition, the sides are all increased or decreased by the same proportion. This is true for any regular polygon or circle.

CHANGE IN AREA WHILE ORIENTATION STAYS THE SAME

- When an image is enlarged or reduced, the area increases/decreases by the square of the scale factor.

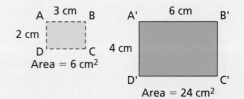

- Notice that the dilatation image has vertices in the same orientation as the original shape—vertices ABCD and A'B'C'D' are clockwise in both.

Tessellations

Tessellations provide an opportunity for students to apply transformations and focus on properties of shapes, especially side lengths and angle measures. Some people use the terms tesselation and tiling interchangeably; while others use tesselation when only one congruent shape is used and tiling when multiple shapes are used.

Relating Tessellations and Transformations

Tessellations involve translating, reflecting, and/or rotating a shape or a combination of shapes in order to cover an area without overlapping or gaps.

2-D Shapes That Tesselate

Most students realize that squares or rectangles will tessellate, and many know that regular hexagons will also work because of their experience with square counters or tiles, attribute blocks, and pattern blocks.

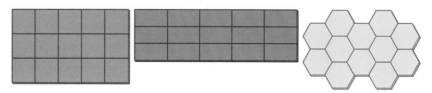

Squares, rectangles, and regular hexagons tesselate

Students may be more surprised to learn that any triangle or quadrilateral can tessellate. This is because the sum of the measures of the interior angles is 180° in a triangle and 360° in a quadrilateral. Shapes will tessellate if the angles that meet at a central point fill a total of 360°.

In the examples below, congruent angles are colour-coded inside each shape to show how the angles combine at a centre point to fill 360°. In the first figure, each angle from the triangle occurs twice in the centre, for a total of 180° × 2, or 360°. In the second figure, all four angles from the trapezoid meet at the centre, for a total of 360°. Side lengths are also important in a tessellation, congruent sides are matched to one another.

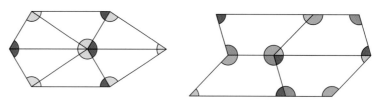

Any triangle or quadrilateral will tile because the angles can be combined to make 360° and congruent sides match.

By exploring regular polygons and trying to tessellate with them, students will discover that the equilateral triangle, the square, and the hexagon work, because their interior angle measures can be combined to fill 360° (6 × 60°, 4 × 90°, and 3 × 120°). Since the interior angle measures of the other regular polygons are not factors of 360°, they will not tesselate surface unless they are combined with other shapes.

Students might also explore what happens when they try to tile with irregular shapes, or when they combine regular and irregular shapes. For example, some irregular pentagons tessellate, depending on their interior angle measures, and regular octagons tessellate when combined with squares.

Teaching Idea | **3.33**

Have students explore the angle properties of regular polygons by trying to tesselate each one. [the equilateral triangle, the square, and the hexagon work]

Focus on BILM 4 by asking: *Where do you see tesselations like that?* [e.g., tiles on bathroom floors and patios]

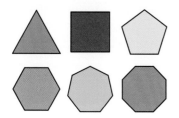

Regular polygons

Teaching Idea | **3.34**

To connect transformations and tesselations and to focus on BILM 2 to 4, ask students to examine an Escher tessellation for transformations. Alternatively, have students examine tessellations created by other students and ask: *Where do you see translations? reflections? rotations?*

Irregular pentagons

Red octagons and a white square

Chapter 4

Measurement

Measurement Fundamentals

Measurement is about assigning a numerical value to an attribute of an object, relative to another object called a unit. A greater measurement implies that one object has "more" of a particular attribute than another.

Measurements are often described in comparative terms, for example, "That package is twice as big as mine," where the implicit unit is the smaller package. Other times, measurements are described in units that are less obvious comparisons, for example, "My package has a volume of 30 cm^3." Here there is an implicit comparison of the package size to the unit 1 cm^3.

Measurements are used to describe attributes of an object in one dimension (length), two dimensions (area), or three dimensions (volume, mass, or capacity). Measurements are also used to describe angles, time, temperature, and money.

There are a number of generic big ideas that underlie any type of measurement at Grades 4 to 8. This chapter differs from the other chapters in this book, in that it begins with a description of these generic big ideas to avoid repetition in each of the individual content areas in the chapter.

BIG IDEAS FOR MEASUREMENT

1. The same object can be described using different measurements.

2. Any measurement can be determined in more than one way.

3. There is always value in estimating a measurement, sometimes because an estimate is all you need or all that is possible, and sometimes because an estimate is a useful check on the reasonableness of a measurement.

4. Familiarity with known benchmark measurements can help you estimate and calculate other measurements.

5. The unit chosen for a measurement affects the numerical value of the measurement; if you use a bigger unit, fewer units are required.

A knowledge of big ideas can help teachers choose, shape, and create tasks and use questioning to help students make powerful connections.

Each teaching idea in this section of the chapter will indicate which Big Idea(s) for Measurement (BIM) can be emphasized. These big ideas will also be referred to, along with the big ideas for each specific content area, in the Teaching Ideas throughout the remainder of the chapter.

6. You can be more precise by using a smaller unit, or by using subdivisions of a larger unit. Also, precision is sometimes limited by the measuring tool that is available.

7. The use of standard measurement units simplifies communication about the size of objects.

8. Measurement formulas allow us to use measurements that are simpler to access in order to calculate measurements that are more difficult to access.

BIG IDEA 1 **The same object can be described using different measurements.**
Any shape has multiple measurement attributes. For a 2-D shape, such as a rectangle, you could measure its dimensions, its area, or the length of its diagonals. For a 3-D shape, such as a prism, you could measure its mass, its volume, its capacity, its surface area, or the length of one of its edges.

Different measurement attributes of an object are not always related, so it is possible for an object that is large in one way to be small in another.

"The yellow bowl is bigger, but the green one is heavier."

Teaching Idea | **4.1**

To focus on BIM 2 and 4, show the shape below and ask: *How does knowing that the grid is 25 square units help you calculate the area of the green shape?* [25 square units − 2 half square units = 24 square units] *How else could you calculate the area?* [e.g., 23 square units + 2 half square units = 24 square units]

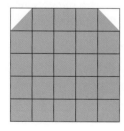

It is important for students to appreciate the value of flexibility in their approach to measuring.

BIG IDEA 2 **Any measurement can be determined in more than one way.**
There are different ways to approach any measurement. Sometimes you measure directly and sometimes indirectly. Sometimes you change the tool you use or how you use the tool. You might choose a different tool to measure the same object, for example, a centimetre ruler versus a millimetre ruler, or you might use the same tool but in a different way.

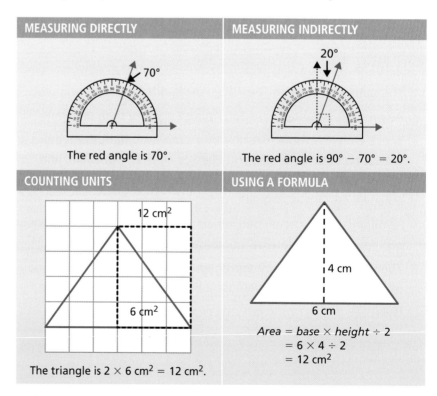

MEASURING DIRECTLY

The red angle is 70°.

MEASURING INDIRECTLY

The red angle is 90° − 70° = 20°.

COUNTING UNITS

12 cm²

6 cm²

The triangle is 2 × 6 cm² = 12 cm².

USING A FORMULA

4 cm

6 cm

Area = base × height ÷ 2
= 6 × 4 ÷ 2
= 12 cm²

BIG IDEA 3 **There is always value in estimating a measurement, sometimes because an estimate is all you need or all that is possible, and sometimes because an estimate is a useful check on the reasonableness of a measurement.**

There are many instances when an estimate is all that is possible, perhaps because of the tools available or the complexity of the situation. More often than not, an estimate is all that is needed in a situation.

AN ESTIMATION SITUATION	ESTIMATING TO CHECK
There are many situations where all that is required to solve a problem or make a decision is an estimate.	It is always a good idea to estimate to check a measurement. Checking estimates often involves using known measurements (Principle 4).

It's 3 Km to the pool. It takes me about 10 min to walk to school and that's about 1 Km. It will probably take us about half an hour to walk to the pool.	The plum's mass is 135 g. That makes sense because the peach was about 300g and the plum is about half as heavy as the peach.

BIG IDEA 4 **Familiarity with known benchmark measurements can help you estimate and calculate other measurements.**

Each time you measure an item with a unit, you are using a benchmark. When you say that an item has a mass of 5 kg, it is only meaningful if you have a sense of how much 1 kg is. This principle also describes situations when you use known measurements to estimate, whether these are "internalized" abstract referents or concrete referents. For example, in **Teaching Idea 4.2**, most students would have an abstract sense of the mass of a pen from past experience and could use that referent and the knowledge that a pen is 20 gibbits to estimate 80 gibbits. In **Teaching Idea 4.1**, one way to calculate the area of the green shape is to compare it to the concrete referent of the 25 cm² grid.

BIG IDEA 5 **The unit chosen for a measurement affects the numerical value of the measurement; if you use a bigger unit, fewer units are required.**

Knowing that the bigger the unit, the smaller the number can help students select appropriate units to measure. For example, it is possible to measure the distance travelled in a car in millimetres or centimetres instead of metres or kilometres but the numerical value would be so large (e.g., 20 000 mm vs. 20 km) that it would be difficult for people to interpret the distance. In the example below, centimetres is probably a more appropriate unit than millimetres; however, if the pencil were slightly longer or shorter than 15 cm, then millimetres would be more appropriate, although the measure would probably be recorded in decimal centimetres, for example, 15.1 cm or 14.9 cm.

Teaching Idea | **4.2**

Tell students that an item has a mass of 80 gibbits (a fictional measure).

To focus on BIM 4 and 7, ask: *Is the item heavy or light? Explain.* [impossible to tell; a gibbet is not a standard unit] *What would help you know?* [knowing how heavy one gibbit is] *My pen has a mass of 20 gibbits. Now do you know if the 80-gibbit item is heavy or light? Explain.* [yes; it won't be heavy since a pen is light and it would only be as heavy as four pens] *Knowing that a pen was 20 gibbits helped you understand 80 gibbits. What other benchmark measurements would have helped you?* [e.g., 40 gibbits or 100 gibbits]

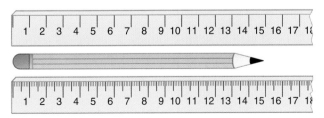

"It's 15 when I measure in centimetres, but 150 when I measure in millimetres."

To focus on BIM 6, have students consider marathon race times. Ensure students know that marathons take many hours to run. Ask: *What unit do you think is used to compare the first and second place finishers?* [seconds or minutes] *Why is it not hours?* [The runners are probably close in time, so you need a more precise unit.] *Why are combinations of hours, minutes, and seconds used to record race times?* [Using only hours would not distinguish the times well enough, and using only seconds would result in numbers that are too big to be meaningful to most people.]

BIG IDEA 6 **You can be more precise by using a smaller unit, or subdivisions of a larger unit. Also, precision is sometimes limited by the measuring tool that is available.**

Help students to see that the reason we use small measurement units like millimetres, milligrams, micrometres, and so on is not only to measure very small objects. We also use small measurements to make comparisons between larger objects that are close in size with respect to the attribute being measured. These smaller units allow for more precision, as would using smaller subdivisions of a larger unit. For example, measuring to the thousandth of a metre is equivalent to measuring in millimetres.

In order to be more precise, students must often change the tool they use. For example, to measure very small objects or compare big objects that are close in size, a metre stick marked in centimetres will not work, but a metre stick marked in millimetres might.

BIG IDEA 7 **The use of standard measurement units simplifies communication about the size of objects.**

If you know an item has a volume of 30 cubes, you have no sense of how big it is because you don't know how big each cube is. If, however, you know it has a volume of 30 cm^3, there is no ambiguity about its volume (although knowing its volume does not indicate its dimensions or shape).

BIG IDEA 8 **Measurement formulas allow us to use measurements that are simpler to access in order to calculate measurements that are more difficult to access.**

For example, consider the problem of finding the area or circumference of a circle. It is always easier to find these attributes by using a formula, because the formula requires only one straight measure (diameter or radius), which is easy to do with a ruler. Calculating curved measures such as area or circumference is much more complicated than measuring it directly, and more prone to error.

Stages of Measurement

For most types of measurement, there are three stages that teachers typically help students move through:

The Definition/Comparison Stage: Students begin to learn to define each measurement, and become aware of and apply a process for comparing items with respect to that measurement, without using units.

The Non-standard Units Stage: Students continue to define the measurement while they learn to measure with non-standard units, such as cheese wedges (for angles) or large linking cubes (for volume).

The Standard Units Stage: Students learn to use measurement tools to measure with standard units, such as millimetres and square centimetres.

Most of the work at the Definition/Comparison Stage and the Non-standard Units Stage will be complete for length, time, mass, and (usually) capacity by the beginning of Grade 4. Often, with temperature, money, and (sometimes) angles, these first two stages are bypassed altogether.

Usually there is some work at the Non-standard Units Stage in both area and volume at Grades 4 to 8. For example, students are introduced to the notion of comparing volumes by counting cubes that are not cubic centimetre cubes (e.g., large linking cubes), or using centimetre cubes but without the knowledge that the cubes are 1 cm^3.

Ask students to build a structure using 12 large linking cubes and a structure using 12 smaller linking cubes.

To focus on BIM 5 and 7, ask: *Why did each of us end up with two structures that have different volumes even though we used 12 cubes both times?* [Each structure used a different size cube.] *If you were to try to copy the large structure using the smaller cubes, what would you need to do?* [I would need a lot more than 12 small cubes.]

The Definition/Comparison Stage

In this stage, students learn what a particular type of measurement is all about by using a comparison procedure (either direct or indirect) to determine which of two items has more of that measure. For example, to compare two angles to see which is greater, students overlap them, lining up one arm to see which angle has a second arm that has been turned more.

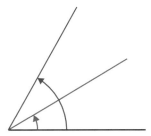

The red angle is bigger than the blue one because it has been turned more.

The Non-standard Units Stage

When familiar objects are used as units for measuring, these are typically referred to as non-standard units since there is no universal meaning that can be attached to them.

Using non-standard units helps to reinforce measurement concepts. For example, with volume, the use of non-standard units reinforces the notion that volume is how much material it takes to build an object.

It is important that non-standard units used be uniform (all the same size and shape) and smaller than the object being measured (so a reasonable degree of precision is possible).

Students should have experience at both the Non-standard Units Stage and the Standard Units Stage for estimating measurements.

The Standard Units Stage

In the Standard Units Stage, students use a standard unit and eventually a variety of standard units to measure. They realize that standard units make it easier to compare and communicate about measurements. For example, if someone says that an object is made with 22 cubes, it is hard to know exactly how big it is. But if they say it is 22 cm^3, the size is well defined.

In Grades 4 to 8, students use a broader variety of standard units than in earlier grades and become familiar with the degree unit for angles.

The Metric System of Measurement

Over 35 years ago, Canada adopted the International System of Units (called SI, for Système International) as its official measurement system. The rationale for adopting the metric system included the fact that this system is already used by most countries around the world—especially in the scientific community—because of its ease of use.

Clearly, one of the attractions of the metric system is its consistency with the place value system. This is particularly valuable when units are renamed, or converted. For example, 3.56 m can easily be renamed as 356 cm without any complicated calculations.

Teaching Idea | **4.6**

Ask students to make these angles:

Angle A: a big angle

Angle B: an angle that is as big as three cheese wedges

Angle C: a 45° angle

To focus on BIM 7, ask: *Why are our angles different for angles A and B, but not for angle C?* [No one was sure exactly what was meant by "big" or by a "cheese wedge," but we all knew exactly what 45° was.]

SI Base Units

While the old imperial system of measurement required Canadians to memorize a wide range of unrelated numbers—12 inches in a foot, 3 feet in a yard, 1760 yards in a mile, and so on—the adoption of the metric system meant that people would simply need to learn a few base units and a series of prefixes that could be applied to these base units.

At the K to 8 level, students are introduced to three of the seven SI base units—the metre, the kilogram, and the second. It may seem odd that the kilogram is called the base unit when the metric prefixes are attached to the word *gram*, not *kilogram*. In fact, the gram was actually a base unit in the older CGS system of measurement; it was superseded by the kilogram, which was considered more convenient for practical purposes. Each of the three base units is based on a precise measurement associated with a science-related situation. While these situations may be of interest to teachers, it is not necessary to teach them, or the term "base unit," to students.

SI Base Units Introduced in Elementary School		
UNIT (SYMBOL)	ATTRIBUTE MEASURED	SYSTÈME INTERNATIONAL DEFINITION
metre (m)	length	1 m is the length of the path travelled by laser-generated light in a vacuum during the time interval of $\frac{1}{299792458}$ of light-speed.
kilogram (kg)	mass	1 kg is a mass equal to the international prototype of one kilogram (about the mass of 1 dm^3 of water at 4 °C).
second (s)	time	1 s is the duration of 9 192 631 770 periods of radiation corresponding to the transition between two hyperfine levels of the ground state of the cesium-133 atom at 0 kelvin.

SI Derived and Customary Units

The International System of Units also uses a number of derived units—units that were created by applying arithmetic operations to the SI base units. Derived-unit measures used in elementary school mathematics include square metres for area, cubic metres for volume, metres per second for speed or velocity, and degrees Celsius for temperature. Other derived-unit measures are not typically dealt with in elementary school. One example is the kilogram per cubic metre, which is used to describe density.

In addition, there are a number of widely used customary units that are considered to be outside the International System of Units, but are accepted for use with SI. These include the litre to describe capacity; the tonne (metric ton) to describe mass; the minute, hour, and day to describe time; and the degree to describe angle arc.

Metric Prefixes

Metric prefixes allow the user to begin with a unit and create larger and smaller units. There are actually 20 prefixes—ten to create larger units and ten to create smaller units—but you commonly use only about half of these.

Metric Prefixes in Common Use

GIGA-	MEGA-	KILO-	HECTO-	DECA-	UNIT	DECI-	CENTI-	MILLI-	MICRO-	NANO-
1 billion units	1 million units	1000 units	100 units	10 units	1 unit	0.1 units	0.01 units	0.001 units	0.000001 units	0.000000001 units

Common Metric Prefixes for Length

KILO-	UNIT	DECI-	CENTI-	MILLI-
1000 m	1 m	0.1 m	0.01 m	0.001 m

Note that the smaller units in the charts above are described with decimals rather than fractions. Decimals are used in the International System of Units because of the relationship of the metric prefixes to the place value system. Using decimals makes it particularly easy to rename a recorded measurement in terms of another unit. For example, to change centimetres to millimetres, you simply need to multiply by ten, which moves each digit one place to the left: 2.4 cm → 24 mm.

At certain grade levels, there must be attention to how to convert from one unit to another. Although many students have no difficulty converting from, say, metres to centimetres, they struggle more with going from the smaller unit to the greater, for example, 15 cm to 0.15 m, and have even more difficulty converting metric area and volume measures, for example, $1 \text{ m}^2 = 10\,000 \text{ cm}^2$ and $1 \text{ m}^3 = 1\,000\,000 \text{ cm}^3$.

Teaching Idea | **4.7**

Ask students how long a "deciday" or a "centiday" would be? [2.4 h; 0.24 h or 15 min]

To focus on BIM 4 and 7, ask: *Why might you not describe a centiday and deciday using the same standard time units?* [Most people have a sense of how long 2.4 h is, but I don't think most people would know how long 0.24 h is, so using minutes is better.] *Why would everyone get the same answers, even though they had never heard of decidays and centidays before?* [The metric system is consistent, so it had to be 24 h divided by 10 and by 100, or their equivalents.]

Length and Area

We use the term length, or linear measure to describe a measurement in one dimension, which includes perimeter. While length describes a one-dimensional attribute of an object, area describes a two-dimensional attribute and standard area units are expressed in terms of length units in two dimensions. Much of the teaching of measurement in elementary school focuses on length and area.

*Each teaching idea in this section of the chapter will indicate which Big Idea(s) for Length and Area (BILA) and/or which Big Idea(s) for Measurement (BIM) from **pages 129 and 130** can be emphasized.*

BIG IDEAS FOR LENGTH AND AREA

1. The length of an object is a one-dimensional attribute, i.e., the number of linear units along a line or curve on the object. Length can be the measurement of a single measure of the object or a combined linear measure, like perimeter.

2. The area of an object is a two-dimensional attribute, i.e., the number of square units it takes to cover a shape on a surface of an object. Area can be a single measure of a 2-D shape on an object or a combined measure of a 3-D shape, like surface area.

Measuring Length

Linear measurements include height, width, length, depth, distance, and perimeter. Ideas about length are introduced even in preschool, probably since estimating and measuring length is something people do often in everyday life. For example, a student might want to know if a pair of pants is long enough, if a distance is too far to walk, or how the heights of family members compare.

Students will have learned a great deal about length in the early grades. At the Grades 4 to 8 level, the focus will be on using more precise units, particularly the millimetre, and possibly other units, like a micrometre. The work will generally be exclusively at the Standard Units Stage of measurement.

If properly instructed, some students may be able to measure in millimetres before they actually realize that the purpose of using millimetres is to gain required precision in a measurement situation.

At some point, students should have an opportunity to think about how many millimetres make up a metre. To do this, they will need to visualize (or see on a metre stick) 100 cm, with 10 mm making up each centimetre, so there are 100×10 mm $= 1000$ mm in all.

There are 10 mm in 1 cm and 100 cm in 1 m,
so there are 100×10 mm $= 1000$ mm in 1 m.

To help reinforce place value concepts, you might also teach units like the micrometre, which is 0.000 001 m (one millionth of a metre) or 0.001 mm (one thousandth of a millimetre).

Teaching Idea | **4.8**

Present two objects that are close in length, such as two pencils.

To focus on BILA 1 and BIM 6, ask: *Why would it be helpful to use millimetres to compare the two pencils?* [Since they are close in length, you'd need a small unit to describe how much longer one is than the other.] *Could you describe the difference using a bigger unit? How?* [Yes, but you'd have to use decimals; e.g., instead of 3 mm, you'd say 0.003 m.] *Why might someone find the measurement in millimetres easier to visualize?* [Visualizing 3 of something is easier than visualizing $\frac{3}{1000}$ of something.]

Renaming Measurements

Part of learning about metric units is learning about the relationships between one unit and another. For example, students need to know that 21.5 cm can be expressed as 215 mm. The ability to rename measurements is useful for

- expressing units in a form that is easier to visualize; for example, a measure of 1.1 m is easier to visualize than 110 cm because it is readily apparent that it is just a little longer than 1 m
- comparing measurement; for example, if two heights are reported in centimetres and metres, respectively, it is easier to compare them if one is renamed
- calculating with measurement formulas; for example, if a rectangle's dimensions are reported in decimetres and centimetres, one of the dimensions has to be renamed to use the area formula
- working with scale; for example, if a map uses 1 cm to represent 1 km, students need to rename 1 km as 100 000 cm in order to understand that the map scale ratio is actually 1 : 100 000

Measuring with Appropriate Precision

As students work with a broader range of units, they will encounter situations where measurements of objects that they wish to compare or measurements of the same object that must be combined through calculation are described with different levels of precision. They need to understand the assumptions they can make about these measurements. For example, if you record a measurement of 3 cm, it is assumed to have been rounded to the nearest whole centimetre, so the measurement could have been anywhere between 2.5 cm and 3.5 cm. If you record a measurement as 3.0 cm, it is assumed to have been rounded to the nearest tenth of a centimetre, so the measurement could have been anywhere between 2.95 cm and 3.05 cm.

To understand the importance of precision, students need to encounter situations where it is important to measure very carefully. There are opportunities in many subject areas, such as art and science, where students need to measure carefully and precisely with small units. Hobbies such as sewing and woodworking can also provide such contexts.

Measuring Perimeter

Although measuring perimeter is often perceived to be distinct from linear measurement, it is really only a variation in which students measure a linear distance that is not a straight line.

MEASURING INDIRECTLY

Initially, students will learn to measure the perimeter of a shape by fitting a string around the shape, cutting it to that length, and then measuring the length of the string. Initially, students may find it more comfortable to refer to the perimeter as the "distance around" the shape, rather than using the more formal term *perimeter*.

Measuring perimeter with a string helps students see that perimeter is still a length measurement—it is the length or distance around an object.

Teaching Idea | **4.9**

Tell students that one desk is reported to be 119 cm wide and another one is 1.2 m wide.

To focus on BILA 1 and BIM 6, ask: *Why can you not be sure which desk is wider?* [119 cm could be anywhere between 118.5 cm and 119.5 cm; 1.2 m could be anywhere between 1.15 m and 1.25 m, which is 115 cm and 125 cm. Therefore, either desk could be wider.]

Teaching Idea | **4.10**

Ask students how many measurements they would need to make in order to calculate the perimeter of each of the following:

- square [1]
- rectangle [2]
- parallelogram [2]
- trapezoid [3]
- equilateral triangle [1]
- isosceles triangle [2]
- scalene triangle [3]

To focus on BILA 1 and BIM 8, ask: *Why is a formula like P = 4s for the perimeter of a square helpful?* [It tells you that you only need to measure one side and then multiply by four to find its perimeter.]

Later, students will measure each side of a shape individually and then combine the side lengths to calculate the perimeter.

Calculating perimeter as shown on the left will inevitably lead to the discovery that there are shortcuts for calculating the perimeter of shapes that have some equal sides, such as rectangles, squares, equilateral triangles, and parallelograms. These shortcuts lead to formulas.

Teaching Idea | **4.11**

Provide string and cardboard circles or use a round waste can. Ask students to figure out the distance around the circle or the base of the can without using the circumference formula.

To focus on BILA 1 and BIM 8, ask: *Why did you have to straighten the string you used to determine its length?* [To measure a length with a ruler, it has to be straight.] *Why would you prefer to use the formula C = πd?* [It is easier and more accurate to measure the diameter and then multiply by π than to measure the circumference because the diameter is a straight length and you can use a ruler to measure it directly.]

Perimeter of a Circle (Circumference)

The perimeter of a circle is called its circumference. The topic of circumference is usually dealt with in Grades 6 and up, although younger students are certainly capable of understanding the notion of measuring around curves.

When the concept of circumference is first introduced, students generally experiment with string or measuring tapes and a variety of circle sizes to discover that the distance around any circle is always about triple the distance across the circle through its centre (the diameter). This presents an opportunity to introduce the concept of π (pi). The circumference is actually π, or 3.141 592 6 ... × the length of the diameter. (If students use a circle with a diameter of 10 cm, they can derive π by dividing the circumference by ten.) Some sources say that this value is named π because, in the Greek alphabet, π is the first letter of the Greek word for perimeter.

It is difficult to be accurate when measuring around a circle unless the circle can actually be cut and straightened out. Therefore, it is helpful that there is a relationship between the diameter and circumference to allow us to calculate the circumference of a circle from its easier-to-measure diameter. Students can multiply the diameter by the approximation 3.14. Or, to get a more precise answer, they can use the π key on a calculator.

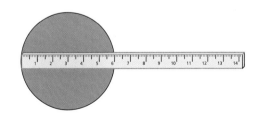

The diameter is 6 cm.   6 [×] [π] [=] 18.8495559

The circumference of the blue circle is about 19 cm.

Perimeter Formulas

Once students have had significant experience working with standard units of length, it will become appropriate to introduce formulas that simplify the calculation of certain linear measures, specifically perimeter. Normally, we help students develop these formulas:

- $P = 4s$ (for squares, where s is the side length)
- $P = 2l + 2w$ (for rectangles, where l and w are the length and width)
- $C = \pi d$ (for circles, where C is the circumference and d is the diameter)
- $C = 2\pi r$ (for circles, where C is the circumference and r is the radius)

You may also introduce other formulas, for example, for the perimeter of an equilateral triangle ($P = 3s$, where s is the side length).

It is important to show when there is flexibility in formulas. For example,

When developing any formula, it is important to help students understand its value—a formula allows us to use a measure that is easy to determine (e.g., side length) to figure out a measure that is more difficult to determine (e.g., perimeter).

DIFFERENT WAYS TO CALCULATE THE PERIMETER OF A RECTANGLE	
$P = 2l + 2w$ (double the length, double the width, and then add)	$P = 2(l + w)$ (add the length and width and then double)

DIFFERENT WAYS TO CALCULATE CIRCUMFERENCE	
$C = d\pi$	$C = 2\pi r$ (since $d = 2r$)

Students should also realize that they can use the formulas "in reverse."

CALCULATING SIDE LENGTH	CALCULATING WIDTH	CALCULATING RADIUS
If you know the perimeter of a square, you can calculate the side length.	If you know the perimeter and the length of a rectangle, you can calculate the width.	If you know the circumference of a circle, you can calculate the radius.
If $P = 4s$, then $s = \frac{1}{4} P$.	If $P = 2l + 2w$, then $w = \frac{1}{2}(P - 2l)$.	If $C = 2\pi r$, then $r = \frac{1}{2}(C \div \pi)$.

Measuring Area

While length describes a one-dimensional attribute of a shape, area describes a two-dimensional attribute—the amount of two-dimensional flat space a shape covers or, in the case of a 3-D shape, the amount of flat space that forms the surface of the shape (surface area). Because of its two-dimensional nature, area is usually expressed in square units, such as square centimetres and square metres.

At the Grades 4 to 8 level, most students will have already been through the Definition/Comparison Stage. It may be appropriate to start with non-standard units with Grade 4 students before moving right to standard metric units.

Area: The Non-standard Units Stage

Non-standard area units are simply handy everyday items of uniform size that can be used to tile a surface. These include sheets of paper or newsprint (to cover big areas), pattern blocks, square tiles, attribute blocks, and ink stamps.

Teaching Idea | **4.12**

Create an outline like the one below, which is made using four hexagon pattern blocks. Ask students to cover it with each type of pattern block to determine the area in terms of each unit:

- yellow [4]   • red [8]
- blue [12]   • green [24]

To focus on BILA 2 and BIM 5 and 7, ask: *Why are the numbers of units different—did the area change?* [no; the smaller the unit, the more units you need to cover the same area]

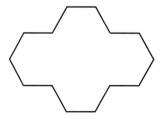

There are many interesting activities for students to explore using non-standard units. For example,

- using sheets of newspaper to compare the areas of two bulletin boards or windows
- using non-standard units to determine the areas of different shapes that can be outlined with the same piece of yarn
- cutting a piece of wrapping paper that is the right size to wrap a gift, and then measuring the area of the paper with non-standard units
- using a box to make a house for a mouse, and then writing a description of the house that includes the floor area, along with features of the home that a mouse might like
- making a paper shape that can be covered by exactly 100 items (100 pennies, 100 cubes, 100 seeds, 100 craft sticks, etc.)
- determining the area of a shape (in non-standard units) in order to share it equally in two parts
- determining the greatest possible area that can be covered with a rolled-out piece of modelling clay

From time to time, it is important for students to have opportunities to make choices about which unit to use to measure an area. These situations help students build their understanding of area concepts and provide concrete experience that will support them later when they move on to work with standard units.

It is important for students to realize that non-standard units are most helpful when the units tile.

Many of the big ideas about measurement can be developed while working with non-standard units, so that students have a fundamental understanding before they begin using standard units; for example, how the size of a unit affects the numerical value of the measurement and that area is a measure of how much flat space it takes to cover a surface of an object.

Using Non-standard Area Units for Developing Area Concepts

USING AREA UNITS THAT DO NOT TILE	USING NON-SQUARE AREA UNITS THAT TILE
Students should measure area with units that do not tile—that is, that do not fit tightly together—only during introductory work with area, when they are exploring why some units work well for measuring area and others do not. When students use units that do not tile, they learn that some of the area is not counted in the measurement; because of this, students can get different results for the same area, depending on how they arrange the units.	Working with non-square units that tile helps students see that squares are not the only shapes that can be used to measure area. Students often have the misconception that area units are called square units because they are in the shape of a square. Using other shaped units can help dispel that.

The area is 10 pennies.

The area is 9 pennies.

Using area units that do not tile can result in different measures of the same area.

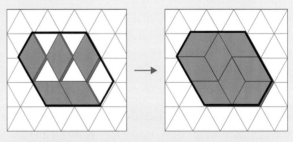

It is important to tile the units when measuring area (i.e., fit them together with no gaps).

Even when units do tile, there are sometimes parts of a shape that cannot be covered with those units. Situations like these help students learn to combine estimating with measuring to get a better estimate of the area. Students need to estimate to add on the parts of an area that cannot be covered with units.

24 whole squares + about 3 half squares is about $25\frac{1}{2}$ squares.

Students can subtract to estimate the area that is not part of the shape.

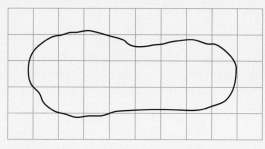

50 squares − about 29 squares is about 21 squares.

Measuring the Same Area with Different Units

Students should have opportunities to measure the same area with different units to help them discover that the area can be described in different ways. It will also help them discover that, as the size of the unit increases, the number of units decreases. For example, an area that covers 49 squares on a centimetre grid could also be covered with only about eight square pattern blocks. Experiences like this also help students discover that it is more efficient to use larger units to measure larger areas and smaller units to measure smaller areas.

Measuring Different Areas with the Same Unit

Not only do students benefit from opportunities to explore area by covering the same shape with different units, they also benefit from using the same unit to measure different areas. A common unit is especially useful when the areas are to be compared.

Measuring Area with a Transparent Grid

Just as a ruler provides a pictorial model for measuring length, a transparent grid provides a pictorial model for measuring area. The grid removes the need for students to either use many individual units or to face the daunting task of iterating the entire space.

A transparent grid can be placed on top of a shape, making it possible to count the number of grid units that cover or partially cover the shape. Like the ruler, the grid allows students to measure different shapes with the same tool. Although a transparent grid usually has square units, there are times when a triangular grid might also be used.

Teaching Idea | **4.13**

Have students use two different-sized grids to measure the area of the same irregular shape, such as a handprint.

To focus on BILA 2 and BIM 6, ask: *Why were more of our answers the same for the smaller grid?* [With big squares, there's more estimating, since it's hard to know what fraction of the square to count. With small squares, there are more squares you're sure of, because they're either in the handprint or they're not.]

Teaching Idea | **4.14**

Sometimes students think that a longer object has a greater area. To focus on how the length of an object is independent of its area, ask students to draw a shape on a grid and then draw another one that is not long but has a greater area.

To focus on BILA 1 and 2 and BIM 1, ask: *Why can't you predict the area of a shape just by knowing how long it is?* [e.g., I can make two shapes that are the same length but one can be wider so it will have a greater area.]

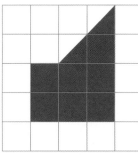

Area = 8 squares   Area = 18 triangles

Transparent grids for measuring area

Measuring Area on a Geoboard

Geoboards, which are dot grids, are useful for exploring areas measured in non-standard units. An interesting approach to calculating the areas of shapes on geoboards is to view the shapes as combinations of rectangles and half-rectangles. For example, to calculate the area of this trapezoid, a student might approach it either way.

Teaching Idea | **4.15**

As students create composite shapes on a geoboard, encourage them to calculate each area in different ways.

To focus on BILA 2 and BIM 2, ask: *Why are there at least two different ways to find the area?* [You can add up the different areas that make up the shape; you can subtract from a larger area that surrounds the shape for the area not covered by the shape.]

COMBINING SHAPES

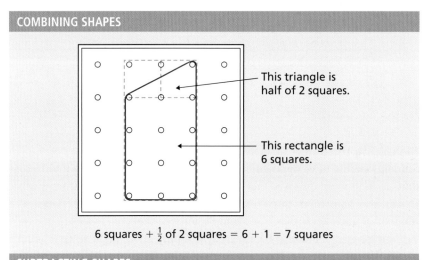

This triangle is half of 2 squares.

This rectangle is 6 squares.

6 squares + $\frac{1}{2}$ of 2 squares = 6 + 1 = 7 squares

SUBTRACTING SHAPES

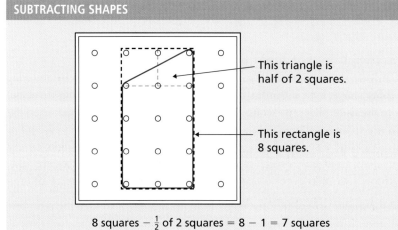

This triangle is half of 2 squares.

This rectangle is 8 squares.

8 squares − $\frac{1}{2}$ of 2 squares = 8 − 1 = 7 squares

Estimating Areas in Non-standard Units

To help students learn to estimate, teachers can provide situations in which a surface is partially measured; the student's task is to estimate how many units it will take to measure the full area. Eventually, students will adopt a similar strategy themselves, using partial measurements as an aid to making reasonable estimates.

Teaching Idea | **4.16**

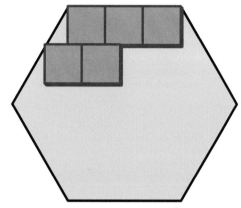

"It takes 5 squares to cover about $\frac{1}{3}$ of the shape, so the hexagon has an area of about 15 squares."

To help students build their understanding of units, it is important for them to have opportunities to measure in one unit, and then use that measurement to make an estimate involving another unit.

Area: The Standard Units Stage

Students can be introduced to standard units when they realize that, rather than measuring areas with non-standard units, which can mean different things to different people, they can use units that everyone will understand.

Introducing Standard Units of Area

The first standard unit of area that students encounter is usually the square centimetre. They then move on to working with square metres, and finally square kilometres. There are also other standard units of area measure, such as the square millimetre and the square decimetre, but these are rarely used.

Although the standard units of area measure are often described in terms of squares with that area, students need to understand that these squares can be dissected and rearranged to form many different shapes, all representing the same area. For example, a shape with an area of one square metre could be a square, but it could also be a triangle, a rectangle, or any other shape that covers the same area. A square metre (or square centimetre or square kilometre) has no defined shape—it is a measure of area and, as a result, is a somewhat abstract notion.

It is important for students to have a firm understanding of the square centimetre, square metre, and square kilometre in terms of how large each unit is, which includes developing referents for each (referents are familiar objects with each area that can be used to help with estimation), how the units relate to each other, and when it is appropriate to use each unit.

Have students cover a business envelope with large sticky notes to measure its area. Then give each group a small sticky note and ask them to estimate how many small sticky notes would be needed to cover the same envelope.

To focus on BILA 2 and BIM 4 and 5, ask: *How did you know the number would be greater for the small sticky note compared to the larger note?* [The unit was smaller, so it takes more of them to cover the same area.] *How did you estimate the number?* [Six small sticky notes cover a big one, so I multiplied the area in large sticky notes by six.]

A square area unit has no defined shape—it is a measure of area and, as a result, is a somewhat abstract notion.

Common Standard Units

THE SQUARE CENTIMETRE (cm²)

Defining a Square Centimetre

A square centimetre is an area equivalent to the area of a square with a side length of 1 cm.

Each shape has an area of 1 cm².

Introducing the Square Centimetre

Initially, square centimetres are introduced in a concrete way, using materials such as centimetre cubes (base ten units), and then pictorially, using centimetre grid paper and transparent centimetre grids. The cubes and grids provide a good transition from non-standard units because students can simply cover a shape with cubes and count to measure the area without focusing on the fact that it is a standard unit. Larger areas can be measured using base ten flats, which cover 100 cm², or by combining flats, rods (10 cm²), and cubes.

Centimetre Referents

Students should develop personal referents for square centimetres to help develop a sense of the unit and to help with estimating. For example, a fingernail or the tip of a baby finger might be a good referent for 1 cm². The palm of a child's hand might serve as a referent for 100 cm².

THE SQUARE METRE (m²)

Defining a Square Metre

A square metre is an area equivalent to the area of a square with a side length of 1 m.

Metre Referents

Students might look for an object in their classroom that covers about 1 m², for example, a small rug, or they might visualize a square metre as an area in which a certain number of students can comfortably stand.

Introducing the Square Metre

The square metre can be modelled with four metre sticks to form the side lengths of a square. The square metre is the space inside. A piece of newspaper or wrapping paper can be cut to fit the space inside, and then it can be used as a transportable square metre unit to help identify objects in the room that are smaller than, about the same size as, or larger than 1 m².

It is important for students to realize that the area can be any shape. They can dissect the square and rearrange the parts to create different shapes, each with an area of 1 m².

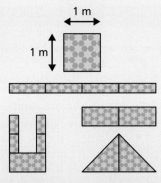

Shapes with an area of 1 m²

THE SQUARE KILOMETRE (km²)

Defining a Square Kilometre

A square kilometre is an area equivalent to the area of a square with a side length of 1 km.

Introducing the Square Kilometre

The only way for students to get a true sense of the size of a square kilometre is to actually go outside and walk around the perimeter of a square that is 1 km². This might include several city blocks. A trundle wheel could be used to measure each side length, or the teacher might have premeasured the lengths using the odometer of a car.

Square Kilometre Referents

This is a unit that is harder for students to get a sense of because of its size. It may be useful for them to associate it with a familiar place.

A square kilometre is about the size of Lakeview Park.

Recording Area Measures in Standard Units

It is important for students to realize that, when they begin to measure with standard units, they need to report measurements in square units. This is different from measuring in non-standard area units, where an area might be reported, for example, as ten blue pattern blocks. Initially, the units might be spelled out, for example, "12 square metres." Eventually, students will be ready to write the units in an abbreviated form using exponents. With standard units, the exponent two is used to indicate that the units have two dimensions. For example, if an area is 12 square centimetres, we write "12 cm^2."

Using a Centimetre Grid

Centimetre grid paper and transparent centimetre grids are useful when students are first beginning to measure area in square centimetres. Like geoboards, centimetre grids are especially useful for dealing with the area of irregular shapes. For example, to estimate the area of the irregular shape below, students might overlay a transparent centimetre grid and then count the number of squares that are completely covered and the number of squares that are mostly covered. They add these together and check to see if the squares with only a little bit covered seem to balance the missing parts in the squares that are mostly covered. They can either increase or decrease their estimate appropriately.

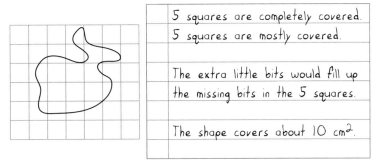

5 squares are completely covered.
5 squares are mostly covered.

The extra little bits would fill up the missing bits in the 5 squares.

The shape covers about 10 cm^2.

The area of an irregular shape can be estimated by counting whole squares and partial squares and then estimating to make adjustments.

Formulas for Calculating Area

During K to 8 instruction in math, students normally have an opportunity to develop and use formulas for determining the areas of rectangles, parallelograms, triangles, circles, and trapezoids.

All too often, formulas are rote procedures students use without understanding for calculating a measurement such as volume. A formula actually describes the shape in terms of relationships among its component measurement attributes. The formula for calculating the area of a rectangle, for example, describes the area in terms of an array of squares:

Area = number of rows × number in each row   or   *Area = length × width*

Students who do not understand the nature of a formula may apply it to the wrong shape or measurement attribute without knowing why this is incorrect, for example, applying the rectangle formula to the side lengths of a parallelogram or confusing area with perimeter (a very common error). As well, students who do not understand will have difficulty rearranging the formula, for example, to calculate the length of a rectangle given the area and the width. The best way to help students understand and recall each formula is to let them play a role in developing the formula.

Teaching Idea | **4.17**

To focus on BILA 2 and BIM 4 and 5, ask: *How would you decide whether to measure the area of a park in square metres or in square kilometres?* [If it were small, I'd use square metres since I know how big 1 m^2 is; if it were big, I would use square kilometres because, if I used square metres, the number would be too big to understand.]

Area of a Rectangle

The first area formula introduced at the elementary level is usually the formula for calculating the area of a rectangle:

Area = length × width   or   *A = l × w*

Using an Array Model The best way to introduce the rectangle area formula is to have students work with rectangles created with square tiles, so that the rectangles are basically shown to be arrays of squares. From earlier work with multiplication and the array meaning or model of multiplication, students will know that, to determine the total number of squares, you multiply the number of rows of squares by the number of squares in each row. In the consolidation discussion, the teacher might ask:

- *What does the length of the rectangle tell you about the number of squares in the rectangle?* (the number of rows of squares)
- *What does the width tell you?* (the number of squares in each row)
- *What does the area tell you?* (the number of squares altogether)
- *Why does it make sense that you can find the area of a rectangle by multiplying the length by the width?*

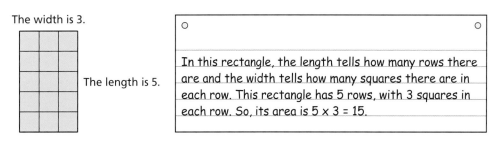

The width is 3.

The length is 5.

The area is 15 squares.

In this rectangle, the length tells how many rows there are and the width tells how many squares there are in each row. This rectangle has 5 rows, with 3 squares in each row. So, its area is 5 x 3 = 15.

Using Relationships Another way to introduce the formula is to have students create some different-sized rectangles on grid paper and determine the length, width, and area of each rectangle by counting grid spaces and grid square side lengths, and then recording the data in a chart.

Teaching Idea | **4.18**

To focus on BILA 2 and BIM 2 and 8, ask: *What tools could you use to measure the area of a rectangle? How would you use each?* [a centimetre grid: you can count the squares; a ruler: you can measure the length and width and multiply] *Which would you prefer to use? Why?* [e.g., the ruler: it's easier and quicker and probably more exact, especially if the rectangle perimeter doesn't fit along the grid lines exactly]

Rectangle	Length (cm)	Width (cm)	Area (cm²)
A	8	2	16
B	5	3	15
C	7	4	28

This chart shows a relationship among area, length, and width: *A = l × w*

Students can then look for relationships in the table among the length, width, and area, which will lead to the formula, *Area = length × width*. They might also notice that they can calculate the length by dividing the area by the width, or the width by dividing the area by the length.

Because the formula for determining the area of a rectangle is usually introduced before any other area formulas, it is a common misconception that the formula *Area = length × width* can be applied to determine the area of shapes other than rectangles. When students develop the formula in a concrete way using arrays and relationships, they are less likely to have this misconception.

Area of a Square

The formula for calculating the area of a square,

Area = side length × side length (A = s × s)

is a special case of the rectangle formula, since *length × width* in this case is the same as *side length × side length*. Students can create different squares on grid paper and record data about the squares in a table such as that shown. They will soon realize that the area of a square is its side length multiplied by itself.

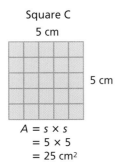

Square C
5 cm

5 cm

A = s × s
  = 5 × 5
  = 25 cm²

Square	Side length (cm)	Area (cm²)
A	8	64
B	7	49
C	5	25

The chart shows that there is a relationship between area and side length: *A = s × s*

Area of a Parallelogram

Once students have a solid understanding of the rectangle formula, they are ready to learn that a related formula—*Area = base length × height*—can be used to calculate the area of a parallelogram. To explore this idea, the teacher can demonstrate that it is possible to transform a parallelogram into a rectangle with the same base length and height by cutting a triangle from one side of the parallelogram and moving it to the opposite side, or by cutting two trapezoids that can be similarly rearranged.

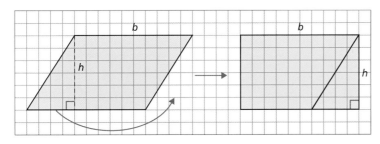

Transforming a parallelogram into a rectangle with the same height (width) and base (length)

Students can then construct and cut out some other parallelograms and test them to see if they can all be transformed the same way. They may notice that the fact that the opposite sides of a parallelogram are equal is the reason why the triangle that is cut off on one side can always be attached to the other side. It is also important for students to recognize that

- the base that is cut remains the same length because the part removed from one end is added to the other end
- the height is measured at right angles to the base
- a single cut along the height forms both sides of the rectangle

For these reasons, the parts of a parallelogram can always be reassembled to form a rectangle with the same base length and height.

Teaching Idea | **4.19**

To help students understand why the area formula for a parallelogram involves its height (altitude) and not the "slanty" side length, have students form a rectangle with cardboard strips (each side a whole number of centimetres). They can place the rectangle on grid paper to measure the area inside and record a sketch of the shape, the base length, the height, and the area in a chart. Students can then change the height to create other parallelograms with the same base.

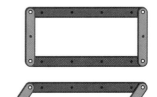

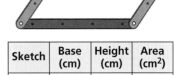

Sketch	Base (cm)	Height (cm)	Area (cm²)
▭	8	5	40
▱	8	3	24

To focus on BILA 2 and BIM 8, ask: *Why does it make sense that the formula uses the vertical height and not the slanty height?* [The base and the "slanty height" stayed the same while the area changed.] *How does knowing the formula make it easier to determine the area?* [It's easy to measure the base length and height with a ruler and then multiply.]

Teaching Idea | 4.20

Ask students to create a parallelogram. Have them determine the area using the area formula but measuring two ways: first using one side as the base and then using the adjacent side as the base (some students will be interested to note that either side of the parallelogram can be used as the base).

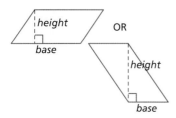

To focus on BILA 2 and BIM 2, ask: *How did you know that the two different approaches would have the same result?* [The area is how much space it covers and the shape did not change, it was just turned.]

In the specific example below, you can see that the area of a parallelogram with a height of 6 cm and a base of 10 cm is the same as the area of a rectangle that is 6 cm wide and 10 cm long.

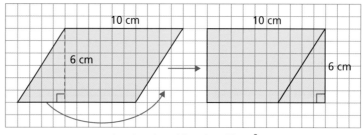

$$A = b \times h = 10 \times 6 = 60 \text{ cm}^2$$

Once students have learned the formula for the area of a parallelogram, they can use it for the area of a rectangle, since a rectangle is just a special parallelogram. For this reason, you will often see the width and length of a rectangle referred to as height and base.

Area of a Triangle

Once students have developed and worked with the parallelogram area formula, the next step is usually the introduction of the formula for determining the area of a triangle:

Area = base × height ÷ 2 or $A = b \times h \div 2$

The triangle area formula is derived from the parallelogram formula and works because every triangle, no matter what type of triangle it is, can be shown to be half of a parallelogram with the same base and height.

Any triangle can be shown to be half of a parallelogram with the same base and height.

The triangle formula can also be written as $A = \frac{1}{2} \times b \times h$ or $A = \frac{b \times h}{2}$, but it makes more sense to use $A = b \times h \div 2$ until students learn to multiply with fractions.

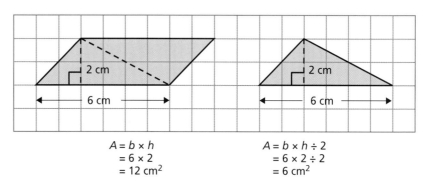

$A = b \times h$
$= 6 \times 2$
$= 12 \text{ cm}^2$

$A = b \times h \div 2$
$= 6 \times 2 \div 2$
$= 6 \text{ cm}^2$

The idea that two triangles with the same base and height must have the same area, even if they look different, is very difficult for some students to understand. This can be modelled on a geoboard, where base lengths and heights can be determined by counting horizontal and vertical units. For example, all three triangles below have the same base (three units) and the same height (four units), so they must also have the same area.

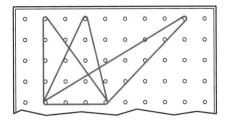

Each triangle has a base length of 3 units and a height of 4 units, so the area of each triangle is 3 × 4 ÷ 2 = 6 square units.

Area of a Trapezoid

A trapezoid can always be viewed as half of a parallelogram. The height of the parallelogram is the same as the height of the trapezoid, and the base length of the parallelogram is equal to the sum of the two bases of the trapezoid. (On a trapezoid, the bases are considered to be the two parallel sides.) The area of any trapezoid can be calculated using the formula $A = h \times (a + b) \div 2$, where h is the height of the trapezoid, and a and b are the lengths of the bases.

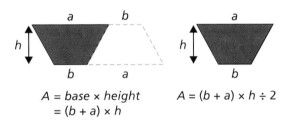

$A = base \times height$
$= (b + a) \times h$

$A = (b + a) \times h \div 2$

Area of Composite Shapes

To calculate areas of composite shapes, students can apply area formulas for more familiar shapes drawn inside or outside the shape in question. For example, the square and triangle areas have been used to calculate the area of this hexagon. Students can figure out the area of the individual parts by counting squares or using a formula.

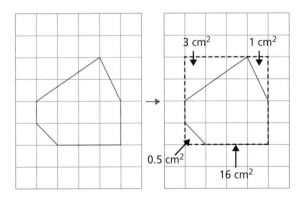

Area of red shape = 16 − (3 + 1 + 0.5) = 11.5 cm²

Teaching Idea | **4.21**

There is a formula for the area of any polygon on square dot paper or a geoboard which is independent of the shape. $A = \frac{B}{2} + I - 1$ describes the area of a shape if B is the number of pegs that a shape's perimeter touches and I is the number of pegs inside the shape.

For example, the shape's perimeter below touches ten pegs ($B = 10$) and has three pegs inside ($I = 3$). Its area is $\frac{10}{2} + 3 - 1 = 7$ units.

To focus on BILA 2 and BIM 2 and 8, ask: *Why is this area formula more useful than the other area formulas?* [It applies to any polygon so you only have to remember one formula.] *When would this formula be less useful?* [when the shape is not on a grid]

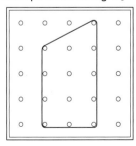

Area of a Circle

A circle can be divided up into many sectors (pie slices) to help students see that these pieces can be formed into an "almost" parallelogram, where the height is the radius of the circle and the base is half the circumference of the circle. The area of the parallelogram is *base × height* or, in this case, *(circumference ÷ 2) × radius*. If students already know that the circumference is equal to $2 \times \pi \times radius$, then they can simplify the formula to $A = \pi \times radius \times radius$.

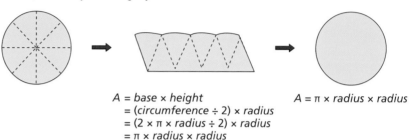

$A = base \times height$
$= (circumference \div 2) \times radius$
$= (2 \times \pi \times radius \div 2) \times radius$
$= \pi \times radius \times radius$

$A = \pi \times radius \times radius$

Area and Perimeter

Students are often surprised to find that shapes with the same area can have different perimeters, and that shapes with the same perimeter can have different areas.

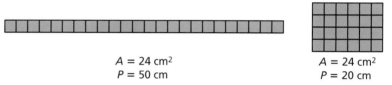

$A = 24$ cm^2
$P = 50$ cm

$A = 24$ cm^2
$P = 20$ cm

Rectangles with the same area can have different perimeters.

Exploring the perimeters of rectangles with the same area can help students make an important discovery—the perimeter of a shape increases as the area is "stretched out" and decreases as the parts are "pushed together" or become more like a circle. The rectangle with the least perimeter for a given area is a square; the shape with the least perimeter for any given area is always a circle.

Surface Area

Students in Grade 6 or 7 and above work with the surface area of 3-D shapes, usually beginning with rectangle-based prisms and then moving on to square-based pyramids and the other prisms and pyramids (focusing first on triangles and then on shapes with regular polygon bases). The surface area of a 3-D shape is the sum of the areas of all of its faces or curved surfaces.

In these grades, students apply the formulas they know for the areas of the individual faces and then combine them rather than use the conventional formula for the surface area of a rectangle-based prism: $SA = 2(lw + lh + wh)$. Note that there is evidence that, if the conventional formula is introduced too early, students may not develop a firm understanding of the concept of surface area. This, of course, applies to the introduction of any formula.

Nets provide an excellent model to work with when calculating the surface area of a 3-D shape because they allow you to visualize all the faces at once and clearly show that surface area is a 2-D measure. From the nets you can see the congruent faces, which can make it easier to find shortcuts for calculating the surface area.

Calculating Surface Area of Prisms

RECTANGLE-BASED PRISM

The rectangle-based prism below has three pairs of congruent rectangle faces.

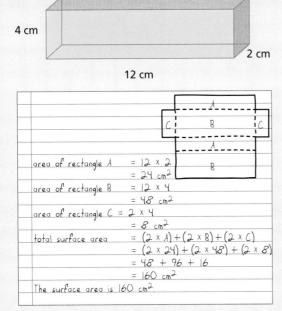

area of rectangle A   = 12 × 2
                       = 24 cm²
area of rectangle B   = 12 × 4
                       = 48 cm²
area of rectangle C = 2 × 4
                    = 8 cm²
total surface area   = (2 × A) + (2 × B) + (2 × C)
                       = (2 × 24) + (2 × 48) + (2 × 8)
                       = 48 + 96 + 16
                       = 160 cm²
The surface area is 160 cm².

- The surface area of a square-based prism would be simpler to calculate since it has four congruent rectangle faces and two congruent square bases.
- A cube has the easiest surface area to calculate because it has six congruent square faces.

TRIANGLE-BASED PRISM

The triangle-based prism below has two congruent triangle bases and three rectangle faces.

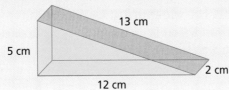

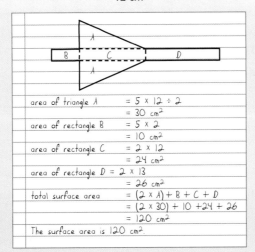

area of triangle A   = 5 × 12 ÷ 2
                       = 30 cm²
area of rectangle B   = 5 × 2
                       = 10 cm²
area of rectangle C   = 2 × 12
                       = 24 cm²
area of rectangle D = 2 × 13
                    = 26 cm²
total surface area   = (2 × A) + B + C + D
                       = (2 × 30) + 10 + 24 + 26
                       = 120 cm²
The surface area is 120 cm².

- The surface area of an equilateral triangle–based prism would be simpler to calculate since it is has three congruent rectangle faces and two congruent triangle bases.

Capacity, Volume, and Mass

Capacity, volume, and mass are all measures that describe attributes of 3-D objects. Capacity units are used to measure the amount of a liquid as well as the size of the inside of an object as if it were a container that could hold liquids. Volume units tell about the size of an object and standard volume units are expressed in terms of length units in three dimensions. Mass units describe the amount of matter in an object.

BIG IDEAS FOR CAPACITY, VOLUME, AND MASS

1. The capacity of an object tells how much it will hold.
2. The volume of a 3-D object tells how much material it takes to build the object.
3. The mass of an object tells how heavy it is.

Measuring Capacity

Most students will have learned a great deal about capacity in the early grades. At the Grades 4 to 8 level, the focus will be on standard capacity units, particularly litres and millilitres.

Although we clearly define capacity as a measure of how much a container could hold, it is important to help students understand why we usually describe the amount of liquids using capacity measures, for example, 250 mL or 1 L of water. Since liquids are kept in containers, and because of their pourable nature, they fill the container's capacity, and it is the capacity of the container that helps us understand the amount of liquid we have. For example, we talk about 1 L of milk when we mean enough milk to fill a 1 L container.

Standard Tools for Measuring Capacity

Just as a ruler helps students learn that they can use a single tool to measure different lengths, a measuring cup illustrates that a single tool can also be used to measure different capacities. A measuring cup is like a ruler in other ways, too. For example, students need to estimate amounts that fall between the marks on the cup, such as 130 mL. They also need to learn to combine full cups with partial ones to measure quantities greater than the capacity of a single measuring cup.

A measuring cup shares many characteristics with a ruler.

With older students, you might investigate standard units like kilolitres, hectolitres, decalitres, or decilitres to broaden students' familiarity with capacity benchmarks and reinforce the base ten place value connection.

Each teaching idea in this section of the chapter will indicate which Big Idea(s) for Capacity, Volume, and Mass (BICVM) and/or which Big Idea(s) for Measurement from pages 129 and 130 can be emphasized.

Teaching Idea | **4.23**

Tell students that a record-size milkshake filled a 22 712 L container.

To focus on BICVM 1 and BIM 6, ask: *Suppose you did not know the capacity of the container. In order to measure it, you poured the milkshake out bit by bit using a 1 kL container (i.e., 1000 L). Would you be able to tell exactly how many litres the milkshake container was? Explain.* [no; the last 1 kL container would only be partly full (712 L out of 1000 L), and you would only be able to estimate] *Suppose the capacity was actually 22 712.2 L. What size container would you need to be able to measure its capacity exactly?* [a 1 dL or 0.1 L container]

Measuring Volume

Students in Grade 4 will generally have a good sense of how to determine whether one object that is obviously bigger than another has a greater volume, so it would make sense to begin at the Non-standard Units Stage rather than the Definition/Comparison Stage.

A common misconception is the confusion between volume and capacity. This is partly promoted by the way volume is sometimes taught. For example, to measure the volume of a prism, we sometimes ask students to fill an assembled net of the prism with cubes and, as a result, end up measuring the capacity using volume units. Perhaps another way to approach a task like this would be to create a model of the prism using cubes instead.

Appropriate Non-standard Units for Volume

Students generally measure volume using uniform units that pack (that is, fit together with no overlap or spaces). The size of the blocks does not matter as long as they are uniform. Examples include centimetre cubes, linking cubes, and building blocks. (Even though centimetre cubes are a standard unit (1 cm³), this has no significance for students at the Non-standard Units Stage, as these students use centimetre cubes as non-standard units.)

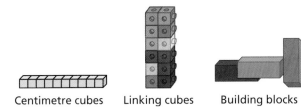

Centimetre cubes   Linking cubes   Building blocks

Uniform items of any size can be used to measure volume in non-standard units.

Rectangle-based prism blocks are often used, but other shapes can work just as well, depending on the object. For example,

"The jar is just a bit smaller than 11 yellow blocks."

Volume and Shape

Students at the Non-standard Units Stage are often surprised to find that shapes that look very different can have the same volume.

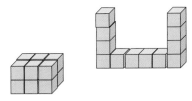

Both structures have a volume of 12 cubes.

To help students explore this concept, a teacher can pose problems like the one in **Teaching Idea 4.25**.

Teaching Idea | **4.24**

Show students two cube structures each made with 20 cubes, both with an empty part that could be filled. One should hold more than the other.

To focus on BICVM 1 and 2 and BIM 1, ask: *What is the volume of each structure?* [20 cubes] *What is the capacity of each (in cubes)?* [red: 12 cubes; green: 4 cubes] *How could you create another structure with the same volume and a different capacity?* [Build a structure with 20 cubes that would hold a different number of cubes, e.g., this structure would hold two cubes.]

Teaching Idea | **4.25**

It is important for students to distinguish between volume and other measurements, such as height.

To focus on BICVM 1 and BIM 1, challenge students to build a shape with a volume of 10 cubes that is taller than another shape with a volume of 12 cubes. For example,

Then ask: *If you know the height of an object, can you predict its volume?* [no; a tall and a short object could have the same volume]

Volume: The Standard Unit Stage

Standard units of volume are related to standard units of length. However, while length is measured in units that are straight lines, volume is measured in cubic units. This is because cubic units represent a space that has three dimensions.

Introducing Standard Units of Volume

Generally, the cubic centimetre is introduced first. This is partly because centimetres are familiar to students, and partly because cubic centimetres are a suitable size for measuring many shapes that are easy for students to handle. In addition, concrete models in the form of centicubes or unit cubes from a set of base ten blocks are readily accessible in many classrooms.

Students should realize that different standard volume units are useful in different situations.

To focus on BICVM 2 and BIM 6, ask: *When would you measure volume in cubic kilometres?* [e.g., the atmosphere] *in cubic millimetres?* [for a very small object, e.g., a strand of hair]

Common Standard Units for Volume

CUBIC CENTIMETRES (cm^3)

Defining a Cubic Centimetre

A cubic centimetre is a volume equivalent to the space occupied by a cube with an edge length of 1 cm.

It is important for students to realize that an object with a volume of 1 cm^3 is not necessarily cube-shaped, as illustrated below.

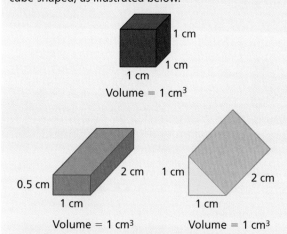

Volume = 1 cm^3

Volume = 1 cm^3 Volume = 1 cm^3

Students can visualize dissecting and reassembling a 1 cm^3 cube to create different shapes with a volume of 1 cm^3.

Cubic Centimetre Referents

Students can refer to centicubes or base ten block unit cubes. A base ten rod serves as a referent for 10 cm^3, a flat for 100 cm^3, and a thousand cube for 1000 cm^3.

CUBIC METRES (m^3)

Defining a Cubic Metre

A cubic metre is a volume equivalent to the space occupied by a cube with an edge length of 1 m.

Students should realize that an object with a volume of 1 m^3 does not have to be cube-shaped. It simply has to occupy the same amount of space as the cube.

Cubic Metre Referents

A good way to model and provide a referent for a cubic metre is to make 12 rolls of newspaper, each 1 m long, and tape them together to form the edges of a cube as shown here.

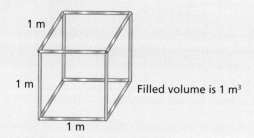

Filled volume is 1 m^3

If this skeleton were solid, its volume would be 1 m^3.

Students can use this model and their estimation skills to look for objects in the classroom and outside that have a greater volume, about the same volume, or a lesser volume. They might also use the model to estimate the volumes of familiar objects, such as a shed or a garbage dumpster.

CUBIC DECIMETRES (dm^3)

The cubic decimetre is another unit of volume that is useful for students in the elementary grades. Its size is suitable for measuring classroom objects, and it can be modelled readily using a thousand cube from a set of base ten blocks.

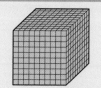

$V = 1$ dm $\times$ 1 dm $\times$ 1 dm
$V = 1$ dm^3

Other standard units of volume include cubic millimetres and cubic kilometres, but these are rarely used in everyday life.

Recording Standard Units of Volume It is important for students to realize that, when they begin to measure volume with standard units, they need to report measurements in cubic units. (This is different from measuring in non-standard units, where a volume might be reported, for example, as 20 blocks or 17 linking cubes.)

Initially, the units might be spelled out, for example, "12 cubic metres." Although this is not formally correct metric usage, it serves to help students build an understanding of the unit before they begin using symbols. Eventually, students will be ready to write the units in an abbreviated form using exponents, for example, 12 cm^3. With standard units, the exponent three is used to indicate that the units have three dimensions.

Relating Units of Volume Students need opportunities to discuss how one cubic unit relates to another. Just as students who are learning about area may be surprised to discover that 1 m^2 = 10 000 cm^2 (and not 100 cm^2), students who are learning about standard units of volume may be even more surprised to find that 1 m^3 = 1 000 000 cm^3.

Relating Linear, Area, and Volume Units

TYPE OF MEASURE	UNIT	EXAMPLE
Linear	1 cm	100 cm
Area	1 cm^2	100 cm, 100 cm, 100 cm 100 cm × 100 cm = 10 000 cm^2
Volume	1 cm^3	100 cm, 100 cm, 100 cm 100 cm × 100 cm × 100 cm = 1 000 000 cm^3

Formulas for Measuring Volume

In K to 6, students deal with formulas, if at all, mainly in the context of working with prisms, especially rectangular prisms. In Grades 7 to 9, they learn formulas for calculating the volumes of other shapes, usually more general prisms, pyramids, cones, cylinders, and spheres. Although volume formulas are officially introduced at the Standard Units Stage, the notion of a formula or shortcut can be introduced at the Non-standard Units Stage.

Volume of a Prism When students build cube models of prisms, many will notice that instead of counting each cube to calculate the volume, they can multiply the number of cubes in each layer (which is represented by the area of the base when dealing with the volume formula at the Standard Units Stage) by the number of layers (which is represented by the height when dealing with the formula).

Teaching Idea | **4.27**

To focus students on BICVM 2 and BIM 8, ask students why you only need two linear measurements to calculate the volume of a cylinder, but three to calculate the volume of a (non-square) rectangular prism. [The width and length of the base of the cylinder are equal (both are the diameter), but that is not true of the prism.]

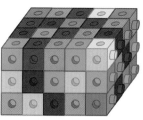

volume = number of cubes in bottom layer
           × number of layers
           = 20 × 3
           = 60 cubes

Tasks like this prepare students for the development of the volume formula.

Volume Formulas for Prisms and Cylinders

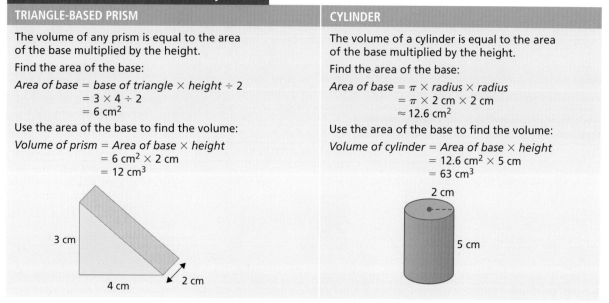

TRIANGLE-BASED PRISM	CYLINDER
The volume of any prism is equal to the area of the base multiplied by the height.	The volume of a cylinder is equal to the area of the base multiplied by the height.

TRIANGLE-BASED PRISM

The volume of any prism is equal to the area of the base multiplied by the height.

Find the area of the base:

Area of base = base of triangle × height ÷ 2
             = 3 × 4 ÷ 2
             = 6 cm²

Use the area of the base to find the volume:

Volume of prism = Area of base × height
              = 6 cm² × 2 cm
              = 12 cm³

3 cm

4 cm   2 cm

CYLINDER

The volume of a cylinder is equal to the area of the base multiplied by the height.

Find the area of the base:

Area of base = π × radius × radius
            = π × 2 cm × 2 cm
            ≈ 12.6 cm²

Use the area of the base to find the volume:

Volume of cylinder = Area of base × height
                = 12.6 cm² × 5 cm
                = 63 cm³

2 cm

5 cm

Volume of Other Shapes Later on, students will learn that they can use the volumes of prisms and cylinders to calculate the volumes of related shapes, such as pyramids, cones, and spheres. Students can first determine concretely how

There are commercial volume relationship kits that can be used to determine these relationships. Students can also work with paper nets of cylinders, pyramids, cones, and rectangle-based prisms.

- the volume of a prism and a pyramid with the same height and base compare by pouring sand from a pyramid container to a prism container with the same height and base. They will discover that they can fill the prism three times.
- the volume of a cylinder and a cone with the same height and base compare by pouring sand from a cone container to a cylinder container with the same height and base. They will discover that they can fill the cylinder three times.
- the volume of a cylinder and a sphere with the same height/diameter compare by pouring sand from a sphere container to a cylinder container with the same height/diameter. They will discover that they can fill the cylinder up to the $\frac{2}{3}$ mark.

The activities above technically compare the capacities of the containers to indirectly compare their volumes. It is important to be explicit about this with the students to reinforce both the distinction and the relationship between volume and capacity.

Once students know how the volumes compare, they can develop formulas for the volume of pyramids, cones, and spheres as shown at the top of the next page. Note that the formulas for a cone and a sphere are introduced later in elementary school.

Relating Volume Formulas

PYRAMID	CONE	SPHERE

PYRAMID

The volume of a pyramid is $\frac{1}{3}$ of the volume of the prism with the same base and height.

$$V_{prism} = \textit{Area of base} \times \textit{height}$$
$$= Bh$$

$$V_{pyramid} = \frac{1}{3} Bh$$

CONE

The volume of a cone is $\frac{1}{3}$ of the volume of the cylinder with the same base and height.

$$V_{cylinder} = \textit{Area of base} \times \textit{height}$$
$$= (\pi \times r \times r) \times h$$
$$= \pi r^2 h$$

$$V_{cone} = \frac{1}{3} \pi r^2 h$$

SPHERE

The volume of a sphere is $\frac{2}{3}$ of the volume of the cylinder that fits exactly around it.

$$V_{cylinder} = \textit{Area of base} \times \textit{height}$$
$$= (\pi \times r \times r) \times 2r$$
$$= 2\pi r^3$$

$$V_{sphere} = \frac{2}{3} \times 2\pi r^3$$
$$= \frac{4}{3} \pi r^3$$

Volume and Surface Area

At the Non-standard Units Stage, students learned that objects with the same volume can have different shapes. Now they can explore the idea that objects with the same volume can also have different surface areas. They will discover that as the shape becomes more like a cube, the surface area decreases.

V = 12 cubes
SA = 50 squares

V = 12 cubes
SA = 40 squares

V = 12 cubes
SA = 32 squares

A short, wide shape has more surface area than a more compact, taller shape with the same volume.

Relating Volume to Capacity through Displacement

When an object is completely immersed in water, it displaces an amount of water that is equal to its volume. Students can compare the volumes of two objects simply by immersing each object in the same amount of water and comparing the changes in the water level. The more the water level rises, the greater the volume of the object.

"The greater an object's volume, the more water it displaces."

Teaching Idea | 4.28

Build two very different cube structures with the same volume but different surface areas (e.g., 1 by 1 by 24 and 3 by 4 by 2), but do not show them to the students. Tell students a structure has a volume of 24 cubes, and ask them to predict the surface area. [You can't predict; e.g., one could have a surface area of 98 squares units and the other 52 square units.]

To focus on BICVM 2 and BIM 1, after showing students the two structures ask: *Does knowing the volume of a shape tell you about its dimensions or its volume?* [no; two shapes can have very different surface areas and dimensions, but the same volume]

Have students immerse a small heavy object and a lighter, larger object in water. Many students are surprised that the lighter object actually displaces more water.

To focus on BICVM 2 and 3 and BIM 1, ask: *What attributes of the object affect the results when measuring volume this way?* [its size by volume] *What attributes do not affect the results?* [mass, width, length, height]

Although displacement is a fairly simple concept, it is usually not introduced until after students have had some experience with measuring length, area, and capacity. Students require a fairly sophisticated understanding of measurement in order to understand the indirect relationship between the object and the displaced water. They also need to understand the relationship between volume and capacity.

Relating Volume and Capacity

At the Standard Units Stage, students are ready to learn that metric units of volume are related not only to one another but also to standard units of capacity. This is why familiarity with units of capacity is a helpful prerequisite for work with volume displacement. When students immerse 1 cm^3 of material in water, it displaces exactly 1 mL of water. When they immerse 1000 cm^3 (1 dm^3), it displaces exactly 1 L. As a result, students can use displacement not only to compare volumes, but also to find the volumes of irregularly shaped objects in standard units.

The water level went from 100 mL to 135 mL, so the volume must be 35 cm^3.

The same relationship (1 cm^3 = 1 mL) can be used to estimate the capacity of a container with a known volume (assuming the sides of the container are not too thick). For example, if the volume of a box is 560 cm^3, then the box has a capacity of just less than 560 mL.

Measuring Mass

In Canada, we generally teach children to use the term *mass*, rather than *weight*, to describe the heaviness or lightness of an object. The differences between mass and weight are listed below:

- Mass measures the amount of matter contained in an object, while weight is a measure of force—the combined effect of mass and gravity.
- The mass of an object is measured by using a balance to compare it to a known amount of matter, but weight is measured on a scale (which uses gravity).
- The mass of an object does not change when location changes, but the weight can change with a change in location (i.e., a change in gravity).

In everyday use, the terms *mass* and *weight* are interchangeable because weight and mass are about the same on Earth; in fact, we often tell students to weigh an object to determine its mass, although some advanced scientific applications require that the distinction be made.

Most students entering Grade 4 will have already dealt with the Definition/Comparison Stage and the Non-standard Units Stage for mass. Likely they will also have already met the kilogram and perhaps the gram as standard units.

There may be attention in Grades 4 to 8 on converting from one mass unit to another using the place value relationships in the metric system (see **page 135**). For example, a student might be asked how to write 219 g in kilograms (since 1 g = 0.001 kg, then 219 g = 0.219 kg).

Measuring Angles

The topic of angles is considered to be part of the geometry strand in some curriculums and part of the measurement strand in others. In some curriculums, certain aspects of the topic of angles are found in both strands. Like volume, angle measurement is generally introduced later than some other types of measurement.

One important reason for learning how to measure angles is that angle comparisons are integral to many geometry concepts. For example, students might measure angles in order to determine whether a triangle is acute or obtuse, whether two shapes are congruent or similar, or whether a four-sided shape has four right angles, making it a rectangle.

Each teaching idea in this section of the chapter will focus on this Big Idea (BIMA) and will indicate which Big Idea for Measurement from **pages 129 and 130** can be emphasized.

BIG IDEA FOR MEASURING ANGLES

1. The measurement of an angle is defined by how much one arm turns around a vertex relative to the other arm.

Measuring Angles

An angle is formed by two line segments or rays with a common endpoint, which is referred to as the vertex of the angle. The line segments or rays are called the arms of the angle.

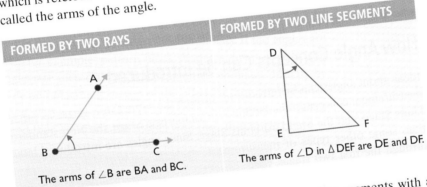

FORMED BY TWO RAYS

The arms of ∠B are BA and BC.

FORMED BY TWO LINE SEGMENTS

The arms of ∠D in △DEF are DE and DF.

Because angles in 2-D are defined as two rays or line segments with a common endpoint, the question is how to measure them. You do not want to use the area inside the angle as a measure, since you want angles to have the same measure whether their arms are long or short.

Two other approaches avoid this problem. One approach is to think of the measure of an angle in terms of what fraction of a circle or full turn it is.

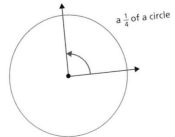

a $\frac{1}{4}$ of a circle

An angle can be thought of as a fraction of a circle or full turn.

Teaching Idea | 4.30

Create two triangles with the same area but a different base angle, such as those below. Ask students whether they think the two angles are the same size or different.

To focus on BIMA 1 and BIM 2, ask: *Why can't you use the area between the arms of an angle to describe its measure?* [You can have the same area but different amounts of turn, or the same amount of turn and different areas.]

the number of values. For example, suppose there are the following numbers of students in six different groups. To calculate the mean group size, you add all six numbers for a total of 30, and then divide by 6 to get 5. The mean indicates that, if all students were redistributed into six equal groups, there would be five students in each group.

WHEN THE MEAN IS AN APPROPRIATE MEASURE	WHEN THE MEAN IS NOT APPROPRIATE
A mean of 5 is an appropriate representation of the data below because there are a few groups that have slightly fewer than 5, a few that have slightly more than 5, and two groups that have exactly 5.	When the group sizes are quite different, but one size is more typical than any other, then the mean is not always the best indicator of the "average" group size. In this case, it might be better to use the mode, which is the number that occurs most often.

Group A	Group B	Group C	Group D	Group E	Group F
3	5	4	7	5	6

The mean is $\frac{3+5+4+7+5+6}{6} = \frac{30}{6} = 5$.

Group A	Group B	Group C	Group D	Group E	Group F
1	1	7	7	7	7

The mean is $\frac{1+1+7+7+7+7}{6} = \frac{30}{6} = 5$.

It is very important early on to establish the notion of the mean with students. This will allow for flexible thinking later.

It is very important early on to establish the notion of the mean with students. This will allow for flexible thinking later with respect to estimating and calculating the mean of a set of data in different ways. It also explains the conventional "formula" for calculating the mean, which students often apply without any understanding—if they understand what the mean is they will understand why the formula requires them to add all the data values and divide by the number of data values.

Working Concretely to Establish the Notion of the Mean
EQUAL GROUPS—ACTING IT OUT

When students are first introduced to the concept of mean, they should have opportunities to act it out and explore it concretely.

For the data set 3, 4, 5, 5, 6, 7, students could

- arrange themselves into six groups: one group of 3, one group of 4, two groups of 5, one group of 6, and one group of 7
- redistribute themselves into six equal groups
- and then count how many students are in each new group.

Students will likely use trial and error to form equal groups.

Another approach would be to form the six groups and then combine to form one large group. They can then divide into six smaller equal groups by moving people one at a time to each group (which models the conventional "formula" $\frac{3 + 5 + 4 + 7 + 5 + 6}{6} = 5$).

"When we made the groups equal, we ended up with 5 in each group, so 5 is the mean."

EQUAL GROUPS—MODELLING WITH CUBES

You can model the same grouping situation using linking cubes. If you use a different colour for each original group, students will see the redistribution of the data in the final result.

The mean number of cubes could be found in two ways:

- combining all the cubes into one larger group, and then sharing them equally among 6 groups (which models the "formula" $\frac{3 + 5 + 4 + 7 + 5 + 6}{6} = 5$)
- making a cube train for each group, as shown here, and then redistributing the cubes until all of the trains are of equal length

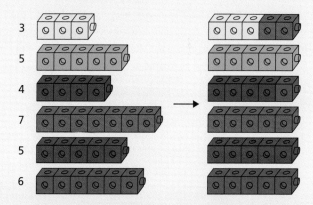

If all the cubes are shared equally among the 6 groups, there are 5 in each group. So the mean is 5.

THE MEAN AS A BALANCE POINT—MODELLING WITH CUBES

Another way to find the mean is to move a ruler or taut string along the cube trains representing the groups until the number of cubes above the mean (to the right of the string) exactly balances the number of empty spaces below the mean (to the left of the string). The vertical line represents the mean, which in the case of the data below is 5. The total length of the arrows on the left of the vertical line, or mean, is 2 + 1 = 3. The total on the right of the mean is 2 + 1 = 3.

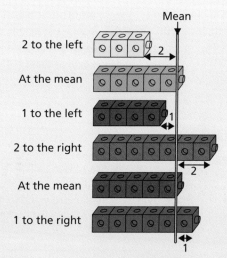

There are 3 cubes to the right of the line and empty spaces for 3 cubes to the left of the line, so the mean must be at the line, which is 5.

Teaching Idea | **5.8**

Ask students to use the picture on the left to show why the mean of 13, 14, 15, 15, 16, 17 is 10 greater than the mean of 3, 4, 5, 5, 6, 7. [e.g., if you add 10 cubes to the left of each row, it won't change the balancing that happens at the other end of the rows, but the value of the line that represents the mean will change from 5 to 15] Then have them create a set of data that would have a mean that is 20 greater. [e.g., add 20 to each data value of 3, 4, 5, 5, 6, 7, so 23, 24, 25, 25, 26, 27]

To focus on BICDD 4, ask: *How does the notion of the mean as a balance make the mean a good way to describe a set of data?* [If it balances the data, it means it's in the middle so it's the single value that is closest to all the data values in the set.]

The Median

The median is another kind of average used to describe a set of data with a single value. It is the middle value when the data points are in order. To find the median group size of the six groups of students below on the left, the values are first put in order from least to greatest. The number in the middle is the median. In this case, there are two middle numbers, but they are the same, so the median is 5. (If the two middle values had been different, the median would have been the mean of the two values.) Notice that half of the numbers are above the median and half are below. Also, note that the median and mean of this set of data are equal. This is not always the case.

It is important to note that the median is usually better than the mean when there are just one or two very extreme data values, as the mean is more influenced by the extreme data than the median. In some cases, it is possible for the median to not be influenced at all.

WHEN THE MEDIAN IS AN APPROPRIATE MEASURE	WHEN THE MEDIAN IS NOT APPROPRIATE
A median of 5 is an appropriate representation of the data below, in the same way that the mean of 5 is appropriate.	A median of 5 is not an appropriate representation of the data below (note that, since the number of values is even, the median is the mean of 1 and 9). The mean (9) might be a more appropriate description.

Group A	Group B	Group C	Group D	Group E	Group F
3	4	5	5	6	7

The median is 5.

Group A	Group B	Group C	Group D	Group E	Group F
1	1	1	9	19	23

The median is $\frac{1+9}{2} = \frac{10}{2} = 5$.

Teaching Idea 5.9

Ask students to create two sets of data with the same mode, one where the mode is an appropriate representation and one where the mode is not appropriate. [e.g., 3, 3, 3, 3, 3, 6 versus 3, 3, 4, 12, 15, 18]

To focus on BICDD 4, ask: *What were you thinking when you made each set of data?* [I made one set where most of the data was the same and one set with very different data values.]

The Mode

The mode is another measure of central tendency used to describe a set of data. It is the most frequent value in the data set. There can be one mode or multiple modes in a set of data. It is important to note that in many situations, there is no mode at all.

If there are many pieces of data, the best way to determine the mode might be to write all the values in order and look for the longest set of matching numbers. It is easy to determine the mode from a bar graph, a line graph, or a stem-and-leaf plot. The mode is the value or values described by the highest bar(s), the highest point(s), or the longest run(s) of identical leaves.

WHEN THE MODE IS AN APPROPRIATE MEASURE	WHEN THE MODE IS NOT APPROPRIATE
A mode of 5 is an appropriate representation of the data below in the same way that the mean and median of 5 are.	When the data values are all quite different from the mode, as in the situation below, using the mode to describe the average group size might be misleading. In this case, the mean (5.7) might be more appropriate.

Group A	Group B	Group C	Group D	Group E	Group F
3	5	4	7	5	6

The mode is 5.

Group A	Group B	Group C	Group D	Group E	Group F
12	12	1	2	3	4

The mode is 12.

Measures of Data Spread

Range

There are many sophisticated measures of the spread or distribution of data. Although mathematicians most frequently use a statistic called a standard deviation, its calculation is difficult. One statistic that is accessible to elementary students is the range. The range is the difference between the least value and the greatest value, so it shows the spread of the data. For example, in the set of values below on the left, the smallest group has 3 students and the largest group has 7. A student would subtract $7 - 3$ to determine a range of 4.

Note that the range is only one measure of the spread. It does not show whether the values are distributed evenly or whether there are gaps or clusters. It also does not indicate the minimum and maximum values. So the range alone does not describe the spread of a set of data very effectively.

Teaching Idea | **5.10**

Ask students to create two sets of data that both have a range of 4. One set should have a mean that is much greater than the other. [e.g., 1, 2, 5 versus 112, 115, 116]

To focus on BICDD 4, ask: *Why is the range alone not a good description of a set of data?* [The range can be the same for a set of low data values and for a set of high data values, so it gives you no idea about the actual data values.]

WHEN THE RANGE DESCRIBES THE SPREAD	WHEN THE RANGE DOES NOT DESCRIBE THE SPREAD
A range of 4 combined with the minimum and maximum values appropriately represents the spread in this set of data.	This set of data has the same range and minimum and maximum values as the set to the left, but the spread is different. A range of 4 gives the group of 7 more weight than is appropriate, since the other groups are all 3.

Group A	Group B	Group C	Group D	Group E	Group F
3	4	5	5	6	7

The maximum value is 7, the minimum value is 3, and the range is 4.

Group A	Group B	Group C	Group D	Group E	Group F
3	3	3	3	3	7

The maximum value is 7, the minimum value is 3, and the range is 4.

Quartiles

Sometimes data is divided into four equal groups called quartiles, not in terms of the intervals or ranges being equal, but so that the number of pieces of data in each group is equal. For example, the data set 3, 4, 7, 8, 10, 12, 16, 18 could be grouped into these four quartiles:

Teaching Idea | **5.11**

Talk to students about the concept of percentiles. Percentiles allow us to be even more specific about data spread than quartiles since they group the data into 100 categories. In a large data set, percentiles let you see the point at which 1% of the population is lower than the data, at which 2% of the population is lower than the data, and so on.

To focus on BICDD 4, ask: *Which percentile values do you think would be most useful to describe the heights of 10-year-old children? Why?* [the 10th, the 50th, and the 90th percentiles; to know if someone is really short, average, or really tall.]

Data set: 3, 4, 7, 8, 10, 12, 16, 18

First quartile	Second quartile	Third quartile	Fourth quartile
3, 4	7, 8	10, 12	16, 18

There are single numerical values attached to the quartiles. Although there are variations in how the quartiles are determined, many calculate them as follows:

- Q_2 is the median of the full set of data
- Q_1 (the lower quartile) is the median of the lower half of the data, not including the median
- Q_3 (the upper quartile) is the median of the upper half of the data, not including the median

For the data set above (3, 4, 7, 8, 10, 12, 16, 18):

- $Q_2 = 9$ (the mean of 8 and 10)
- $Q_1 = 5.5$ (the mean of 4 and 7)
- $Q_3 = 14$ (the mean of 12 and 16)

First quartile	Second quartile	Third quartile	Fourth quartile
3, 4	7, 8	10, 12	16, 18

Q_1 — 5.5 Q_2 — 9 Q_3 — 14

Early Ways to Make a Circle Graph (continued)

USE A FRACTION CIRCLE

Initially, students can use fraction circles that match their data (e.g., a circle divided in tenths to represent 10 people), and simply colour the appropriate fractions of the circle.

For example, if 5 out of 10 students in a group have one sibling, then $\frac{5}{10}$ of the circle would represent that group.

How Many Siblings We Have

3 siblings

2 siblings

0 siblings

1 sibling

USE A PERCENT CIRCLE

Once students understand the concept of percent, they can graph data for groups of 100 or percent data using a percent circle (a fraction circle with a circumference divided into hundredths).

For example, if 7 out of 33 students wear glasses, then 7 out of 33 or about 21% or $\frac{21}{100}$ of the circle would represent these students. (When rounding percents, you may need to adjust the numbers to make sure the total is exactly 100%.)

Who Wears Glasses?

Glasses

No glasses

USE A HUNDREDTH GRID

A variation of the circle graph made from a percent circle is a hundredth grid. Students count the squares on a grid to construct or read the graph.

A hundredth grid is similar to a circle graph because, like a circle graph, it shows the relationships of the parts to the whole, as well as the relationships among the parts.

For example, this grid shows that 50% of the members of the team were on the team last year, 33% were on a different team last year, and 17% are new to the sport.

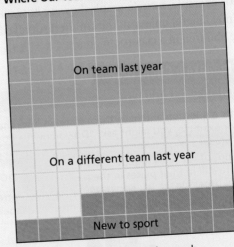

Where Our Team Members Played Last Year

On team last year

On a different team last year

New to sport

A variation of a circle graph

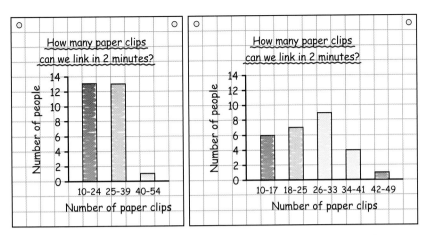

Different intervals for the same data can make the data look very different.

Double Bar Graphs

Sometimes it is useful to look at two sets of data simultaneously, for example, to compare the number of sisters that classmates have with the number of brothers. To show how two different sets of data are alike or different, it would be appropriate to create a double bar graph. A legend is used to help the reader differentiate the two sets of data.

The Number of Brothers and Sisters We Have

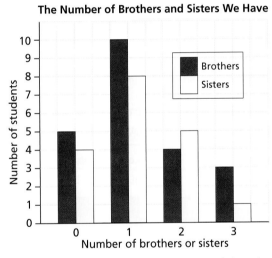

A double bar graph allows you to compare two sets of data simultaneously.

IMPORTANT POINTS ABOUT ...

BAR GRAPHS

1. A grid square is used to represent the same quantity throughout.
2. Bars should be separated to indicate that they represent discrete data, although it is not incorrect if there are no spaces between the bars.
3. If there are no pieces of data for a category or interval, a space can be left where the bar would be, although it is not required.
4. Bar graphs can be vertical or horizontal.
5. Both axes should be labelled. Each bar should have a label, which might be a discrete topic category, a discrete number category, or a numerical interval category. The other axis, the scale axis, is labelled numerically.
6. Axes headings should be used for clarity, as necessary.
7. It is important to include a concise but meaningful title to help the reader understand the graph.

Teaching Idea | **5.14**

To focus on BIDAD 2 and 7, show students the graph below and ask: *Is it easy to tell how many people were surveyed?* [no; unless you know that all university graduates are also high school graduates, then it's not obvious that 150 people were surveyed] *What is the potential problem with the categories?* [They allow for cross-classification—100 people are in both groups.]

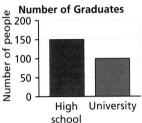

Teaching Idea | **5.15**

Students can investigate the variations of double bar graphs that computer programs create. Sometimes the two related pieces of data are shown by one vertical bar in two colours that consists of one bar stacked on top of the other rather than two bars side by side.

To focus on BIDAD 2, 6, and 7, ask: *Which display do you think is most effective to compare the pairs of related values?* [e.g., the side-by-side bars make it easier to see which values are greater] *Why do you think the other format is used?* [e.g., it makes it easier to see the totals in the categories]

Results of Flipping Two Coins 100 Times	
Outcome	Number of times
2 heads	||||| ||||| ||||| ||||| ||
2 tails	||||| ||||| ||||| ||||| ||||| ||
1 head and 1 tail	||||| ||||| ||||| ||||| ||||| ||||| ||||| ||||| ||||| |||||	

Flipping a head and a tail appears to be most likely.

Students can then be asked to explain why the 1 head/1 tail outcome occurred more often than predicted. They might conclude that the sample size was too small and, if they add more trials, the results for the three outcomes will even out. If they continue flipping, however, they will soon discover this is not the case. At this point, students need to begin looking at theoretical probability, which requires analyzing the situation.

There are different models students can use to do this. A tree diagram, for example, can be used to systematically display the possible results for flipping two coins. They will discover that there are, in fact, 4 possible outcomes because there are 2 outcomes "hidden" in the 1 head/1 tail outcome (1 head/1 tail and 1 tail/1 head). A new experiment with all 4 outcomes listed might show a more equal distribution of results, as shown below.

Results of Flipping Two Coins 100 Times	
Outcome	Number of times
2 heads	||||| ||||| ||||| |||||	
2 tails	||||| ||||| ||||| ||||| |||||	
1 head and then 1 tail	||||| ||||| ||||| ||||| ||||| ||||| ||
1 tail and then 1 head	||||| ||||| ||||| |||||	

There are 4 possible outcomes when flipping two coins.

Using Logic and Analysis to Make Predictions

Experimental probability is determined by the results of an experiment that has already occurred. In contrast, theoretical probability involves analyzing possible outcomes in advance and using logic and reason to predict what is likely to happen. For example, when you roll a die, you expect that if the die is fair, each outcome is just as likely as any other. Since there are 6 possible outcomes when rolling a die, the probability of rolling a particular number is $\frac{1}{6}$.

Teaching Idea | **5.31**

Have students predict the probability of spinning a 1 on this spinner:

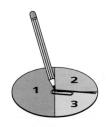

To focus on BIPr 2, ask: *Since there are 3 sections on the spinner, is the probability $\frac{1}{3}$?* [no; the 3 outcomes are not equally likely] *How could you redraw the spinner to make the probability more obvious?* [see spinner below]

Like experimental probability, theoretical probability can also be used to determine the likelihood of an event that involves more than one outcome. For example, for the spinner shown below, a student could expect to spin a number less than 3 half the time, since the sectors for numbers less than 3 cover $\frac{1}{2}$ the spinner's area.

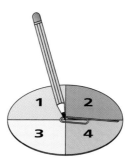

The probability of spinning a number less than 3 is $\frac{1}{2}$.

Reconciling Theoretical and Experimental Probability

It is important for students to understand that while theoretical probabilities can tell them what is likely to happen, they cannot be certain about what will happen in a particular situation. Many students feel that something must be wrong if they know intuitively that the probability of flipping heads is $\frac{1}{2}$, but they have just flipped 10 times and ended up with 7 heads, or even more.

Some of the results that surprise students might not be those that teachers expect. For example, some students think that it is harder to roll a 6 on a die than the lower numbers. It is only through many experiences that they will overcome these misconceptions.

This is where teachers should emphasize the importance of a reasonably large sample size. The results of a small number of trials or just one experiment can be misleading, but when multiple trials are used and/or an experiment is repeated many more times, the experimental probability will gradually approach the theoretical probability. This seeming incongruity between experimental and theoretical probability can sometimes be confusing for students.

Determining Theoretical Probability

When all the possible outcomes of an event are equally likely, the theoretical probability of the event can be expressed as a fraction, as shown below:

$$\text{Theoretical probability} = \frac{\text{Number of favourable outcomes}}{\text{Total number of outcomes}}$$

For example, the probability of rolling an even number with a single roll of a fair die is $\frac{3}{6}$ because there are 3 equally likely favourable outcomes (2, 4, and 6), and 6 equally likely possible outcomes (1, 2, 3, 4, 5, and 6).

It may be worthwhile talking about what the word *favourable* means; it can be confusing. For example, suppose you want the probability that you will not get a 4 when you roll a die. In this case, the favourable outcome is the unfavourable situation of not getting a 4, that is, the favourable outcomes are 1, 2, 3, 5, and 6.

Determining theoretical probability becomes more challenging when the outcomes are not equally likely. For example, the spinner on the left at the top of the next page has 4 sectors, 3 of which are red, but the probability of spinning red on this spinner is not $\frac{3}{4}$. The sectors are different sizes, and so

Teaching Idea | **5.32**

Students often enjoy the game aspect of probability. Have them play this game several times to see if it is a fair game.

Two students take turns rolling two dice. If the product is even, player A gets a point. If the product is greater than 7, player B gets a point.

To focus on BIPr 2 and 4, ask: *How could you figure out if the game is fair?* [experimentally: play it many times, or theoretically: make a list of all the outcomes for each player and determine the total number of possible products (27 out of 36 possible products are even and 22 out of 36 possible products are greater than 7) and compare (they will discover that the game is not fair)]

Note that a 6 by 6 multiplication table is a handy tool for determining the theoretical probability of different products.

they are not equally likely. To determine the probability, the sectors need to be partitioned into equal parts as shown below on the right.

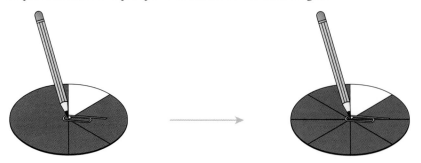

The probability of spinning red is not $\frac{3}{4}$, even though 3 out of 4 sectors are red.

The probability of spinning red is $\frac{7}{8}$, since 7 out of 8 *equal* sectors are red.

Models for Determining Theoretical Probability

There are several models that students can use to determine theoretical probability.

Tree Diagram

A tree diagram is a graphic organizer that can help students determine all the possible outcomes for a compound event.

The tree diagram below shows the possible results of 2 coin flips. The outcomes for each coin (heads/tails) are equally likely. By looking at the tree diagram, you can draw conclusions about the theoretical probability. For example, the diagram not only shows that there are 4 possible outcomes, but also that

- the probability of flipping 2 heads in a row (HH) is 1 out of 4, or $\frac{1}{4}$
- the probability of flipping 2 tails in a row (TT) is 1 out of 4, or $\frac{1}{4}$
- the probability of flipping 1 head and 1 tail (HT or TH) is 2 out of 4, or $\frac{2}{4}$

Possible Outcomes of
Flipping a Coin Twice

Flip 1	Flip 2	Outcomes
Head	Head	HH
	Tail	HT
Tail	Head	TH
	Tail	TT

There are 2 outcomes, heads or tails, for Flip 1. For each of those, there are 2 outcomes for Flip 2. So there are 4 possible outcomes altogether.

It is important for students to make sure that any outcomes represented in the diagram are equally likely, or the results will be misleading. In the coin example above, flipping a head is as likely as flipping a tail. Consider the situation in **Teaching Idea 5.33**. Because red takes up $\frac{2}{3}$ of the area of the spinner and blue takes up $\frac{1}{3}$, spinning red is twice as likely as spinning blue. When creating the tree diagram for this situation, this can be modelled by including red twice and blue once both times. The tree diagram clearly shows that there are nine possible outcomes and four of them are red and blue (either red, then blue [RB] or blue, then red [BR]). Note that the tree diagram also shows that the probability of spinning red and blue in a particular order is $\frac{2}{9}$.

Area Model

Another model for displaying and/or determining the theoretical probability of a compound event is an area model. Again, this pictorial model relies on students' abilities to represent the situation with equal parts. Suppose you want to show the possible results of flipping a coin twice. The first diagram clearly shows that the coin will land on heads half the time and tails the other half. The second diagram shows the possible combinations for 2 flips, making it very clear in a visual way that the probability of flipping 2 tails is $\frac{1}{4}$.

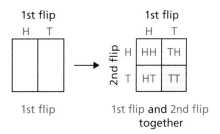

The probability of flipping 2 tails in 2 flips is $\frac{1}{4}$.

Area models can also be used to model probability situations where the outcomes are not equally likely. For the spinner situation described in **Teaching Idea 5.33**, the first spin could be modelled by a square divided vertically into $\frac{2}{3}$ red and $\frac{1}{3}$ blue, and then the second spin would be modelled by dividing the square horizontally into $\frac{2}{3}$ red and $\frac{1}{3}$ blue. The result would be a square that was $\frac{1}{9}$ BB, $\frac{4}{9}$ RR, $\frac{2}{9}$ RB, and $\frac{2}{9}$ BR.

Simulation Experiments

A simulation is an experiment that is used to determine or estimate the experimental probability of a situation that would be difficult or impossible to determine the theoretical probability for through analysis. Simulations are designed to mimic or model the conditions of the problem.

Simulation experiments combine aspects of both experimental and theoretical probability. They are experiments that use probability devices that are chosen because of their theoretical probability. For example, if you want to simulate something that has a $\frac{1}{2}$ chance of happening, you might use a coin because the theoretical probability of flipping a head is $\frac{1}{2}$.

Litter Simulation Experiment

Suppose students want to determine the probability that there will be at least one male and one female cub in a litter of three snow leopard cubs.

> **STEP 1 FIND A SUITABLE RANDOM DEVICE.**
>
> Since male and female cubs are about equally likely to be born, students could use a two-sided counter (yellow and red) to model each birth. A yellow flip could represent the birth of a male cub, and a red flip the birth of a female. (A coin, a two-part spinner, or a draw bag with 5 cubes of one colour and 5 cubes of another colour would also be suitable devices, because each has 2 equally likely outcomes that have a theoretical probability of $\frac{1}{2}$.)
>
> Until they are developmentally ready, students may struggle with how flipping counters can really provide information about the gender of baby snow leopards. It is critical to spend the time talking about the mathematical similarity of the situations, recognizing, of course, the very different contexts. In some ways, it is no different from recognizing that the expression 3 + 2 is just as valid for determining the total number of 3 elephants and 2 elephants as 3 counters and 2 counters.

Teaching Idea | **5.34**

Ask students to create a simulation to solve this problem. Alison, Kaley, and Jen are best friends. They want to be in the same homeroom in Grade 7. There are four Grade 7 homerooms and two Grade 7/8 homerooms. All the classes will have the same number of students. The Grade 7/8 homerooms are half Grade 7 students. Students are randomly assigned to homerooms. What is the probability that all three girls will end up in the same homeroom?

To focus on BIPr 5, ask: *How did you design a device to model the problem situation?* [I made a spinner with 5 sections and divided the fifth section into two equal parts. That made sure the class sizes were represented correctly.] *Describe your experiment.* [I spun the spinner three times, the first spin was to see what class Alison ended up in, the second spin was for Kaley, and the third spin was for Jen. I did this 50 times and then counted how many times all three girls ended up in the same class. I used that number as the numerator for the denominator of 50.]

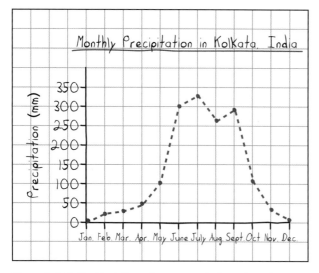

One might infer that the rainy season is from June to September in Kolkata.

It is very important for students to justify any inferences or conclusions that they draw from the data.

Scatter plots and line graphs are particularly well suited for inferring a relationship between variables. For example, the line graph below shows running times for Olympic winners for the 100-m women's race from 1948 to 2004.

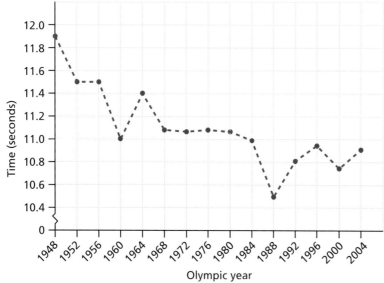

You can infer from this graph that there is a relationship between the passage of time and sprinting speed; that is, the times to complete the race have been improving over time. However, this improvement has not been continuous, since there is a slight backslide starting in 1992, although 1988 might be considered an anomaly and ignored. As well, you can also predict that, as the times get closer to 10 s, the rate of decrease will get less and less.

A valuable activity for developing students' ability to infer is to provide a graph without labels and challenge them to infer what the graph might be about (see **Teaching Idea 5.23**).

Teaching Idea | **5.23**

Provide the untitled and unlabelled graphs below and ask students to think of a set of data that might be represented by each graph [e.g., bar graph: how many students in a class have a certain number of siblings; circle graph: what fraction of students in the class have each hair colour]

To focus on BIDAD 2, 3, 6, and 7, ask: *Why did you say "How many ..." for the bar graph but "What fraction ..." for the circle graph?*

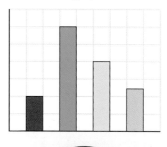

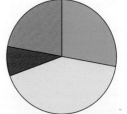

Interpolation and Extrapolation

You can make inferences from a graph by interpolating and extrapolating. Interpolation happens when the reader makes a judgment about values between two given or known values on a graph or in a table. Extrapolation is about predicting values beyond the known values using trends in the data. As with other inferences, it is important to give students a chance to justify their interpolations and extrapolations.

It is important to give students a chance to justify their interpolations and extrapolations.

EXAMPLE GRAPH

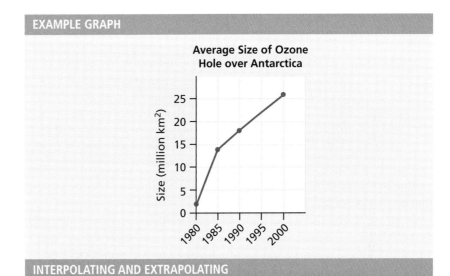

INTERPOLATING AND EXTRAPOLATING

This line graph above shows that the size of the ozone hole has been increasing since 1980, with the sharpest increase from 1980 to 1985.

- No data was plotted for 1995, but students might interpolate that the size of the hole in 1995 was between 18 million km^2 and 26 million km^2— perhaps about 22 million km^2.
- Students might also notice that the ozone hole increased by roughly 4 million km^2 every five years until 2000, and predict that this increase will continue. Students would then extrapolate that the size of the hole will be 30 million km^2 in 2005, and 34 million km^2 by 2010.

Generally, students are less comfortable with extrapolation than with interpolation and this is reasonable. It is important to emphasize that when line graphs display discrete data, interpolation is inappropriate because there are no data points between the known data points. Many line graphs do not use dashed lines even if the data is discrete, so students should be cautioned to analyze line graphs carefully before making any assumptions about interpolating.

Misleading Graphs

One of the main points students need to understand is that it is important to be careful when they are creating and interpreting visual displays. Graphs can be misleading, sometimes deliberately, but more often accidentally. The following types of errors can lead to misleading graphs.

Misuse of Intervals

Interval bar graphs are supposed to use intervals of equal size. If unequal intervals are used, the graph can be misleading. For example, perhaps in an attempt to determine which interval to use to display a set of data points, a student has created two graphs. The graphs show the same data, but tell very different stories.

Teaching Idea | **5.24**

To focus students on BIDAD 2 and 7, tell students that there were people of these ages at a party: 2, 2, 4, 5, 8, 9, 10, 12, 17, 25, 30, 36, 37, 38, 40. Ask: *How might you create intervals for a bar graph to show how many people were at the party to fairly reflect the range of ages?* [0–4, 5–9, 10–14, …] *How would you create intervals to give a misleading impression of the age range?* [1–34 and 35–69]

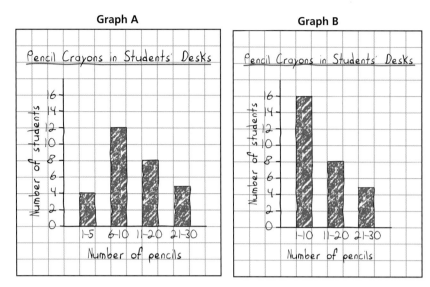

Using different intervals changes the way the data looks.

Graph B is accurate, indicating that the group of students who have 1 to 10 pencil crayons is double the size of the group with 11 to 20, and more than triple the size of the group with 21 to 30. Graph A minimizes these differences because the data from the first interval on Graph B is graphed as two intervals on Graph A.

Misuse of Scale

The importance of choosing an appropriate scale is exemplified by the two graphs below. In Graph A, a scale of 100 is used, necessitating the use of an axis break (shown by the squiggle in the axis). In Graph B, the same data is shown using a scale of 1000. The impressions left by the graphs showing the same data are quite different.

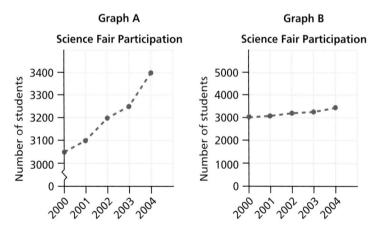

The use of a different scale and an axis break can make a graph misleading.

Teaching Idea | **5.25**

To focus on BIDAD 2 and 7, provide students with data about hot lunch choices: 90 students chose 2 slices of pizza, 80 students chose 3 slices of pizza, and 60 students chose 1 slice of pizza. Ask: *How would you graph the data to make it seem like a lot more students chose 2 slices of pizza than 1 slice?* [Start the scale at 5 using an axis break from 0 to 5, and use intervals of 20.]

In Graph C below, cheddar appears to be the most popular cheese by far because the height of the cheddar bar relative to the other two bars appears so much greater. In fact, this is only because of the axis break in the vertical scale between 0 and 950 people. As a result, the first square in each bar on Graph C actually represents 950 students, while the other squares only represent 5 people. In Graph D, every square represents 100 students. Graph D leaves a more accurate impression of students' cheese preferences, indicating that about the same number of people chose each type of cheese.

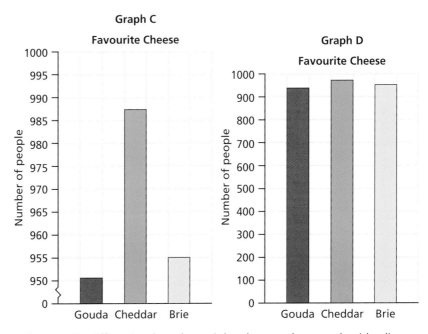

The use of a different scale and an axis break can make a graph misleading.

Probability

Data Management and Probability are usually considered one strand in mathematics because it is only through collecting, organizing, representing, and analyzing data that students are able to draw conclusions about probability.

BIG IDEAS FOR PROBABILITY

1. An experimental probability is based on past events and experiments. It is calculated by dividing the number of favourable trials by the total number of trials and is always a value from 0 to 1.

2. A theoretical probability is based on an analysis of what could happen. It is calculated by dividing the number of favourable outcomes by the total number of equally likely outcomes and is always a value from 0 to 1.

3. In a probability situation, you can never be sure what will happen next.

4. An experimental probability approaches a theoretical probability when enough random samples are used.

5. Sometimes a probability can be estimated by using an appropriate model and conducting an experiment.

Each teaching idea in this section of the chapter will indicate which Big Idea(s) for Probability (BIPr) can be emphasized.

Introducing Probability

Probability is the study of measures of likelihood for various events or situations. How likely it is to rain tomorrow, how likely it is that a contestant will spin a particular number on a game show, and how likely it is that a particular candidate will win an election are all examples of probability situations.

Probabilities are sometimes calculated theoretically. For example, the probability of tossing a head on a coin is $\frac{1}{2}$ since there are two equally likely possible outcomes (heads and tails), and only one of the two is favourable (heads). Experimental probability is the probability you calculate using the results of an actual experiment.

Unlike most mathematical situations, which are based on predictability and patterns, in probability situations you can never be sure what will happen on a particular occasion, unless the event is either impossible or certain. For example, many students will predict that if they flip a coin, say, 10 times, it will land heads up 5 times. Although this is a good prediction, it is still quite possible to flip the coin 10 times and have it land heads up anywhere from 0 to 10 times. This is because of the randomness of flipping a coin. If you repeat the 10-flip experiment many times, the average number of heads in 10 flips is likely to get closer and closer to 5.

Unlike most mathematical situations, in probability situations you can never be sure what will happen on a particular occasion.

Tell students that you flipped a coin 12 times and got these results: HTHHTHHTHHTH.

To focus on BIPr 3, ask: *What do you think will happen next?* [can't be sure] *Why might the pattern not continue?* [Each flip, no matter what has happened before, is just as likely to be tails as heads.] Repeat for the result TTTTTTTTTTTT to address the common misconception that the next flip is more likely to be H because so many tails have been flipped (or T, because that appears to be the pattern).

Probability Misconceptions

Young students and even some adults often have naive ideas about probability. These ideas are often common misconceptions. For example, many students think that if a coin is flipped and lands heads up 5 times in a row, it is likely to land heads up again. Others think just the opposite—that it is unlikely to land heads up next time. It takes experience and/or logic to understand that the chance that the coin will land heads up each time is no different than it was the time before. Each time, the probability of landing heads up is $\frac{1}{2}$. This is because each flip of the coin is independent of the next flip.

Outcomes, Events, and Sample Space

The possible results of a probability experiment are called outcomes. For example, the two equally likely possible outcomes of flipping a coin are heads and tails. If you roll a standard die, the six equally likely possible outcomes are $1, 2, 3, 4, 5,$ and 6. You could be interested in the probability of any one of those outcomes, such as rolling a 5, or you could be interested in the probability of a combination of outcomes, such as rolling an even number (which includes the outcomes $2, 4,$ and 6). The term *event* is used to describe the outcome or combination of outcomes that you are interested in, such as the event of rolling a 5 or the event of rolling an even number.

Students will eventually encounter the term *sample space* to describe the set of all possible outcomes. They also learn the term *favourable outcome* to describe the outcome or combination of outcomes that they are "looking for." For example, if students are conducting an experiment to calculate the probability of the event rolling a number greater than 3, the sample space is any roll from 1 to 6, and any roll of 4, 5, or 6 is a favourable outcome.

A SIMPLE EVENT

A simple event is simply one event, such as the probability of drawing a heart from a deck of cards.

To determine the experimental probability of drawing a heart, a card can be drawn from a deck of cards numerous times (with the card returned to the deck each time and the deck shuffled), and the number of favourable outcomes can be compared to the total number of trials. Thus, if hearts are drawn 23 times in 100 draws, the experimental probability is $\frac{23}{100}$.

Since the last draw is a heart, the experimental probability of drawing a heart is $\frac{23}{100}$.

The theoretical probability is calculated by analyzing the deck of cards. Since there are 52 cards and 13 are hearts, then the theoretical probability is $\frac{13}{52}$ or $\frac{1}{4}$.

A compound event is an event involving two or more simple events, such as drawing a heart and then another heart from a deck of cards.

To determine the experimental probability, two cards can be drawn from a deck of cards (returning the first card and shuffling the deck before drawing the second card each time), and the number of favourable outcomes compared to the total number of trials. If two hearts are drawn in a row 9 times in 100 trials, the experimental probability is $\frac{9}{100}$.

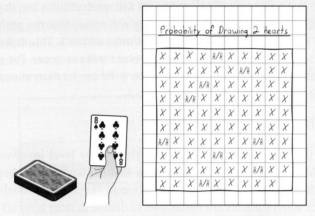

Since the last draw is not a heart, the experimental probability of drawing two hearts is $\frac{9}{100}$.

The theoretical probability for this, which could be determined using a tree diagram or area model, is $\frac{1}{16}$. (It could also be calculated by multiplying the probability of drawing a heart each time, $\frac{1}{4} \times \frac{1}{4} = \frac{1}{16}$, but this is beyond elementary school mathematics.)

Compound Events: Independent and Dependent

In the card-draw experiments described on the previous page and above, the first card you draw has no bearing on the second. When one event does not influence the other, the two events are said to be independent events. If, on the other hand, the first card had not been returned to the deck before the second draw, then the two events would be dependent events, meaning that the second event is influenced by the first. For example, if the first draw is a heart, then there is less chance of drawing a heart on the second draw because one heart has already been removed from the deck. Although at the elementary level it is certainly not necessary to use the language of independent and dependent events, the concept will come up and can be dealt with informally at the middle school or junior high level.

Probabilities as Fractions

Young students describe probability using words such as unlikely, very likely, not very likely, and so on. As students get older, they assign fractions to probabilities. The least possible value of a probability is 0, which indicates that the event could never occur (impossible); the greatest is 1, which indicates that the event must always occur (certain). In between are fractions that describe unlikely events (such as $\frac{1}{4}$) and likely events (such as $\frac{3}{4}$).

Probability as a value is defined as the fraction or ratio that relates all favourable outcomes to all possible outcomes. For example, an experimental probability of $\frac{3}{10}$ means that, out of 10 trials, a favourable outcome occurred

Making Predictions Based on Experimental Results

Making predictions is an important part of any experiment. You want students to use prior experiences to help them predict what will happen next. Students need to learn that it is all right if their predictions do not come true; in fact, it is critical to their understanding of the concept of probability that they recognize that they can never be certain of what will happen. Still, students should be able to explain or justify the predictions they make. For example, after a die-rolling experiment to determine the probability of rolling a 6 (as shown at the bottom of the previous page), a teacher might ask the class to predict how many 6s they expect to see in another 10 rolls of the die. Many students will say 2, basing their prediction on the prior results, but other students may think that since 3 did not come up in the last set of rolls, there will be more 3s next time and, therefore, fewer 6s. As students develop their understanding of probability, they will begin to use theoretical probability to make their predictions. (See *Theoretical Probability* below.) Once they understand that the theoretical probability of rolling a 6 is $\frac{1}{6}$, because there are 6 possible equal outcomes, they might reason that there will be one or two 6s in the next 10 rolls.

Experimental results are used even in older grades for problems that are difficult to model theoretically. For example, to determine the probability of getting a run of six heads when you flip a coin, the theoretical math is beyond students' abilities and an experimental approach makes more sense.

Theoretical Probability

Students usually start to think about theoretical or expected probability more formally somewhere around Grade 5 or 6 after they have had ample opportunity to work with experimental probability. However, informal work with theoretical probability begins much earlier. For example, when students look at a spinner that is $\frac{1}{2}$ red and predict they will spin red half the time, or when they predict that about half of their coin flips will land heads up, they are using theoretical probability, but informally.

Making the Transition from Experimental to Theoretical

Coin-flipping experiments serve as an excellent lead-in to the notion of theoretical probability. For example, students are asked to predict what will happen if they flip two coins 100 times. In an initial discussion of the possible outcomes, they might decide that there are three possible outcomes: 2 heads, 2 tails, and 1 head/1 tail. Since a coin is a fair device (the two outcomes are equally likely, therefore the device has no bias), it might appear as if each of the three possible outcomes will occur about the same number of times. When students conduct the experiment with a large enough sample size, they will discover the three outcomes are not equally likely, as shown on the next page.

Teaching Idea | 5.30

To bridge from experimental to theoretical probability, ask students to make a spinner that they predict will spin green 6 times in 10 spins. Or ask them which spinner below is more likely to result in 6 green spins in 10 spins. [Spinner A]

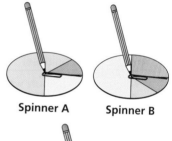

Spinner A   **Spinner B**

Spinner C

To focus on BIPr 4, ask: *Why might you want to try the experiment more than once or twice to be sure you created or chose the right spinner?* [you can never be sure what will happen, but you can be more sure when you do a lot of spins]

Early Ways to Make a Circle Graph (continued)

USE A FRACTION CIRCLE

Initially, students can use fraction circles that match their data (e.g., a circle divided in tenths to represent 10 people), and simply colour the appropriate fractions of the circle.

For example, if 5 out of 10 students in a group have one sibling, then $\frac{5}{10}$ of the circle would represent that group.

USE A PERCENT CIRCLE

Once students understand the concept of percent, they can graph data for groups of 100 or percent data using a percent circle (a fraction circle with a circumference divided into hundredths).

For example, if 7 out of 33 students wear glasses, then 7 out of 33 or about 21% or $\frac{21}{100}$ of the circle would represent these students. (When rounding percents, you may need to adjust the numbers to make sure the total is exactly 100%.)

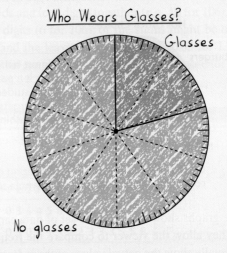

USE A HUNDREDTH GRID

A variation of the circle graph made from a percent circle is a hundredth grid. Students count the squares on a grid to construct or read the graph.

A hundredth grid is similar to a circle graph because, like a circle graph, it shows the relationships of the parts to the whole, as well as the relationships among the parts.

For example, this grid shows that 50% of the members of the team were on the team last year, 33% were on a different team last year, and 17% are new to the sport.

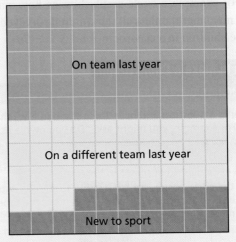

A variation of a circle graph

A student flips 3 two-sided counters to simulate the birth of 3 cubs in a litter. If 3 reds or 3 yellows come up, the student records a check mark beside "no" to indicate that only male cubs or only female cubs were born. However, if a combination of reds and yellows comes up, the student records a check mark beside "yes" to indicate that the cubs were of different genders.

Trials are done repeatedly to create a sample large enough to use as a basis for drawing conclusions. The results of 20 trials might be as shown in the chart below.

Trials are done repeatedly to create a sample large enough to use as a basis for drawing conclusions.

My Flips

Trial	1	2	3	4	5	6	7	8	9	10	11	12	13	14	15	16	17	18	19	20
Yes	✓	✓		✓	✓	✓	✓	✓				✓	✓	✓	✓	✓	✓	✓	✓	✓
No			✓						✓	✓									✓	

The results in the chart above show that the experimental probability of having at least 1 male cub and 1 female cub in a litter of 3 was $\frac{16}{20}$ or $\frac{4}{5}$, which means the probability is very likely.

It is important for students to consider the results and realize that

- there could be different results in another experiment with 20 trials
- the more trials in an experiment, the more confident you can be about making generalizations that will apply to the greater population

Fishpond Simulation Experiment

Another common simulation is for students to estimate the number of times they would need to go "fishing" in a fishpond to collect three different prizes that are equally represented.

The fishpond contains equal numbers of fish labelled A, B, and C. Each letter represents a different prize. A spinner with 3 equal sections would make an appropriate device, since each of the 3 prizes is equally likely. Each section of the spinner could correspond to a different prize. (Again, any random device with 3 equally likely outcomes would be suitable. For example, a die could be used and rolling a 1 or 2 could be prize A, rolling a 3 or 4 prize B, and rolling a 5 or 6 prize C.)

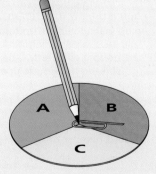

A random device with three equally likely outcomes

(continued)

Spinning simulates catching a fish and winning a certain prize. Students spin the spinner as many times as necessary until they have spun each letter at least once. That is one trial.

The chart below shows the results of one possible experiment with 15 trials. A circled letter indicates when all three prizes have been caught and the trial is over.

Spinning for Fish

Trial \ Spins	1	2	3	4	5	6	7	8
1	A	B	A	Ⓒ				
2	C	A	Ⓑ					
3	A	A	A	B	A	Ⓒ		
4	C	C	A	Ⓑ				
5	C	A	C	Ⓑ				
6	A	B	A	Ⓒ				
7	C	C	C	C	B	Ⓐ		
8	B	C	C	Ⓐ				
9	C	A	C	A	A	Ⓑ		
10	A	B	A	B	A	A	A	Ⓒ
11	C	B	C	C	Ⓐ			
12	A	C	A	Ⓑ				
13	B	A	A	A	Ⓒ			
14	A	B	A	A	Ⓒ			
15	B	B	A	B	B	Ⓒ		

The least number of spins it took to get all 3 letters was 3 and the greatest was 8.

From the chart, students might conclude any of the following:

To collect all three prizes:

- You would need to catch between 3 and 8 fish (the minimum and maximum).
- Most people would need to catch about 4 fish (the mode).
- Most people would need to catch about 5 fish (the mean or the median).

Notice how statistics can help with summarizing the results and drawing conclusions.

In an experiment such as this, students will clearly see the connection between data and probability.

It is valuable to have students consider data they have collected. For example, after the experiment, you might ask questions such as the following:

- In which trials did it take 6 spins to collect all 3 prizes? [3, 7, 9, and 15]
- How can you tell that 3 was the least number of spins required? [Trial 2 was the shortest and it had 3 trials.]
- Which trial took the greatest number of spins? How many spins did it take? [trial 10; 8 spins]
- Why do you think you did not get results like the ones in Trial 10 very often? [the actual probability is close to 5]

References

Abrahamson, D., and Cigan, C. (2003). A design for ratio and proportion instruction. *Mathematics Teaching in the Middle School*, 9, 493–501.

Aldrich, F., and Sheppard, L. (2000). Graphicacy: The fourth 'r'? *Primary Science Review*, 64, 8–11.

Ash, R. (Published yearly). *The top 10 of everything: Canadian edition*. Toronto: Dorling Kindersley.

Ash, R. (1996). *Incredible comparisons*. Toronto: Dorling Kindersley.

Attia, T.L. (2003). Using school lunches to study proportion. *Mathematics Teaching in the Middle School*, 9, 17–21.

Baek, J.M. (1998). Children's invented algorithms for multidigit multiplication problems. In Morrow, L., and Kenney, M.J. (Eds.). *The teaching and learning of algorithms in school mathematics*. Reston, VA: National Council of Teachers of Mathematics, 151–160.

Baek, J.M. (2005). Research, reflection, practice: Children's mathematical understandings and invented strategies for multi-digit multiplication. *Teaching Children Mathematics*, 12, 242–247.

Battista, M.T. (1999a). Fifth graders' enumeration of cubes in 3D arrays: Conceptual progress. *Journal for Research in Mathematics Education*, 30, 417–448.

Battista, M.T. (1999b). The importance of spatial structuring in geometric reasoning. *Teaching Children Mathematics*, 6, 170–177.

Battista, M.T. (2003). Understanding students' thinking about area and volume measurement. In Clement, D., and Bright, G. (Eds.). *Learning and teaching measurement*. Reston, VA: National Council of Teachers of Mathematics, 122–142.

Bay-Williams, J.M., and Martinie, S.L. (2003). Thinking rationally about number and operations in the middle school. *Mathematics Teaching in the Middle School*, 8, 282–287.

Beckman, C.E., Thompson, D.R., and Austin, R.A. (2004). Exploring proportional reasoning through movies and literature. *Mathematics Teaching in the Middle School*, 9, 256–262.

Billings, E.M.H. (2001). Problems that encourage proportional sense. *Mathematics Teaching in the Middle School*, 7, 10–14.

Bloomer, A. (1997). *Getting into area: Grades 3–6*. Palo Alto, CA: Dale Seymour Publications.

Borko, H., and Putnam, R. (1995). Expanding a teacher's knowledge base: Cognitive psychological perspective on professional development. In Guskey, T., and Huberman, M. (Eds). *Professional development in education: New paradigms and practices*. New York: Teacher's College Press, 35–65.

Bremigan, E.G. (2003). Developing a meaningful understanding of the mean. *Mathematics Teaching in the Middle School*, 9, 22–27.

Bright, G.W., Brewer, W., McCain, K., and Mooney, E.S. (2003). *Navigating through data analysis in grades 6–8*. Reston, VA: National Council of Teachers of Mathematics.

Bright, G.W., and Hoeffner, K. (1993). Measurement, probability, statistics, and graphing. In Owens, D.T. (Ed.). *Research ideas for the classroom: Middle grades mathematics*. Reston, VA: National Council of Teachers of Mathematics, 78–98.

Chapin, S., Koziol, A., MacPherson, J., and Rezba, C. (2003). *Navigating through data analysis and probability in grades 3–5*. Reston, VA: National Council of Teachers of Mathematics.

Clements, D.H., and Battista, M.T. (1992). Geometry and spatial reasoning. In Grouws, D.A. (Ed.). *Handbook of research on mathematics teaching and learning*. New York: Macmillan, 420–464.

Cooke, M.B. (1993). A videotaping project to explore the multiplication of integers. *Arithmetic Teacher*, 41, 170–171.

Cramer, K., Wyberg, T. and Leavitt, S. (2008). The role of representations in fraction addition and subtraction. *Mathematics Teaching in the Middle School*, 13, 490–496.

Cuevas, G.J., and Yeatts, K. (2001). *Navigating through algebra, grades 3–5*. Reston, VA: National Council of Teachers of Mathematics.

Empson, S.B. (2001). Equal sharing and the roots of fraction equivalence. *Teaching Children Mathematics*, 7, 421–425.

Fennema, E. and Romberg, T.A. (Eds.). (1999). *Mathematics classrooms that promote understanding*. Mahwah, NJ: Lawrence Eerlbaum Associates.

Friel, S., Rachlin, S., and Doyle, D. (2001). *Navigating through algebra, grades 6–8*. Reston, VA: National Council of Teachers of Mathematics.

Fuson, K.C. (1990). Conceptual structures for multiunit numbers: Implications for learning and teaching multidigit addition, subtraction, and place value. *Cognition and Instruction*, 7, 343–403.

Fuson, K.C., (2003). Toward computational fluency in multidigit multiplication and division. *Teaching Children Mathematics*, 9, 300–305.

Gavin, M.K., Belkin, L.P., Spinelli, A.M., and St. Marie, J. (2001). *Navigating through geometry in grades 3–5*. Reston, VA: National Council of Teachers of Mathematics.

Greenes, C., and Findell, C. (1999). Developing students' algebraic reasoning abilities. In Stiff, L.V., and Curcio, F.R. (Eds.). *Developing mathematical reasoning in Grades K-12*. Reston, VA: National Council of Teachers of Mathematics, 127–137.

Greenes, C., and Findell, C. (1998). *Groundworks: Algebra puzzles and problems, grades 4–7*. Chicago: Creative Publication.

Groth, R.E. (2007). Reflections on a research-inspired lesson about the fairness of dice. *Mathematics Teaching in the Middle School*, 13, 237–243.

Harper, S.R. (2004). Students' interpretation of misleading graphs. *Mathematics Teaching in the Middle School*, 9, 340–343.

Hedges, M., Huinker, D., and Steinmeyer, M. (2005). Supporting teacher learning: Unpacking division to build teachers' mathematical knowledge. *Teaching Children Mathematics*, 11, 478–484.

Heller, P., Post, T., Behr, M., and Lesh, R., (1990). Qualitative and numerical reasoning about fractions and rates by seventh- and eighth-grade students. *Journal for Research in Mathematics Education*, 21, 388–402.

Huinker, D. (1999). Letting fractional algorithms emerge through problem solving. In Morrow, L., and Kenny, M.J. (Eds.). *The teaching and learning of algorithms in school mathematics*. Reston, VA: National Council of Teachers of Mathematics, 170–182.

Jones, G.A., and Thornton, C.A. (1993). *Data, chance & probability, grades 4–6 activity book*. Lincolnshire, IL: Learning Resources.

Kent, L.B. (2000). Connecting integers to meaningful contexts. *Mathematics Teaching in the Middle School*, 6, 62–66.

Knuth, E.J., Alibali, M.W., Hattikudur, S., McNeil, N.M., and Stephens, A.C. (2008). The importance of equal sign understanding in the middle grades. *Mathematics Teaching in the Middle School*, 13, 514–519.

Lamb, L.C., and Thanheiser, E. (2006). Understanding integers: Using balloons and weights software. In Alatorre, S., Cortina, J.L., Saiz, M., and Mendez, A. (Eds.). *Proceedings of the 28th Annual Meeting of the North American Chapter of the International Group for the Psychology of Mathematics Education*, 2, 163–164.

Lamon, S.J. (1999). *Teaching fractions and ratios for understanding: Essential content knowledge and intructional strategies for teachers* (2nd ed.). Mahwah, NJ: Lawrence Erlbaum Associates.

Langrall, C.W., and Swafford, J. (2000). Three balloons for two dollars: Developing proportional reasoning. *Mathematics Teaching in the Middle School*, 6, 254–261.

Lannin, J., Townsend, B.R., Armer, N., Green, S., and Schneider, J. (2008). Developing meaning for algebraic symbols: Possibilities and pitfalls. *Mathematics Teaching in the Middle School*, 13, 478–483.

Lesh, R., Post, T., and Behr, M. (1988). Proportional reasoning. In Hiebert, J., and Behr, M. (Eds.). *Number concepts and operations in the middle grades*. Reston, VA: Lawrence Eerlbaum and National Council of Teachers of Mathematics, 93–118.

Li, Y. (2008). What do students need to learn about division of fractions? *Mathematics Teaching in the Middle School*, 13, 546–552.

Linchevski, L. and Williams, J. (1999). Using intuition from everyday life in "filling" the gap in children's extensions of their number concept to include the negative numbers. *Educational Studies in Mathematics*, 39, 131–147.

Mack, N.K. (1995). Confounding whole-number and fraction concepts when building on informal knowledge. *Journal for Research in Mathematics Education*, 26, 422–441.

Mack, N.K. (1998). Building a foundation for understanding the multiplication of fractions. *Teaching Children Mathematics*, 5, 34–38.

Mack, N.K. (2004). Connecting to develop computational fluency with fractions. *Teaching Children Mathematics*, 11, 226–232.

Martinie, S.L., and Bay-Williams, J.M. (2003). Using literature to engage students in proportional reasoning. *Mathematics Teaching in the Middle School*, 9, 142–147.

McCoy, L., Buckner, S., and Munley, J. (2007). Probability games from diverse cultures. *Mathematics Teaching in the Middle School*, 12, 394–402.

Miller, J.L., and Fey, J.T. (2000). Proportional reasoning. *Mathematics Teaching in the Middle School*, 5, 310–313.

Mokros, J., and Russell, S.J. (1995). Children's concept of average and representativeness. *Journal for Research in Mathematics Education*, 26, 20–39.

Oppenheimer, L., and Hunting, O. (1999). Relating fractions and decimals: Listening to students talk. *Mathematics Teaching in the Middle School*, 4, 318–321.

Pagni, D.L. (2005). Angles, time and proportion. *Mathematics Teaching in the Middle School*, 10, 436–441.

Parker, M., and Leinhardt, G. (1995). Percent: A privileged proportion. *Review of Educational Research*, 65, 421–481.

Perlwitz, M.D. (2005). Dividing fractions: Reconciling self-generated solutions with algorithmic answers. *Mathematics Teaching in the Middle School*, 10, 278–283.

Pugalee, D.K., Frykholm, J., Johnson, A., Slovin, H., Malloy, C., and Preston, R. (2002). *Navigating through geometry in grades 6–8*. Reston, VA: National Council of Teachers of Mathematics.

Reeves, C.A., and Webb, D. (2004). Balloons on the rise: A problem solving introduction to integers. *Mathematics Teaching in the Middle School*, 9, 476–482.

Schifter, D., Bastable, V., and Russell, S.I. (1997). Attention to mathematical thinking: Teaching to the big ideas. In Friel, S., and Bright, G., (Eds.). *Reflecting on our work: NSF teacher enhancement in mathematics K–6.* Washington, DC: University Press of America, 225–261.

Sharp, J., and Adams, B. (2002). Children's constructions of knowledge for fraction division after solving realistic problems. *The Journal of Educational Research*, 95, 333–347.

Shaughnessey, J.M. (2006). Research on students' understanding of some big concepts in statistics. In Burrill, G.F., and Elliott, P.C. (Eds.). *Thinking and reasoning with data and chance.* Reston, VA: National Council of Teachers of Mathematics, 77–98.

Small, M. (2006). *PRIME: Data management and probability: Background and strategies.* Toronto: Thomson Nelson.

Small, M. (2007). *PRIME: Geometry: Background and strategies.* Toronto: Thomson Nelson.

Small, M. (pre-press). *PRIME: Measurement.* Toronto: Nelson Education Ltd.

Small, M. (2005a). *PRIME: Number and operations: Background and strategies.* Toronto: Thomson Nelson.

Small, M. (2005b). *PRIME: Patterns and algebra: Background and strategies.* Toronto: Thomson Nelson.

Stephan, M., Cobb, P., Gravemeijer, K., and Estes, B. (2001). The role of tools in supporting students' development of measuring conceptions. In Cuoco, A.A., and Curcio, F.R. (Eds.). *The roles of representation in school mathematics.* Reston, VA: National Council of Teachers of Mathematics, 63–76.

Streefland, L. (1996). Negative numbers: Reflections of a learning researcher. *Journal of Mathematical Behavior*, 15, 57–77.

Thompson, C.S., and Bush, W.S. (2003). Improving middle school teachers' reasoning about proportional reasoning. *Mathematics Teaching in the Middle School*, 8, 398–403.

Tripathi, P.N. (2008). Developing mathematical understanding through multiple representations. *Mathematics Teaching in the Middle School*, 13, 438–445.

Tzur, R. (1999). An integrated study of children's construction of improper fractions and the teacher's role in promoting learning. *Journal for Research in Mathematics Education*, 30, 390–416.

Warrington, M.A., and Kamii, C.K. (1998). Multiplication with fractions: A Piagetian constructivist approach. *Mathematics Teaching in the Middle School*, 3, 339–343.

Whiteley, W. (2006). Exploring the parallelogram through symmetry [Online]. http://dynamicgeometry.com/general_resources/user_groups/jmm_2006/download/ExploringParallelograms.doc.

Whiteley, W. (2004). Visualization in mathematics: Claims and questions toward a research program [Online]. http://www.math.yorku.ca/Who/Faculty/Whiteley/Visualization.pdf. Cited 2008 Aug 08.

Wiggins, G., and McTighe, J. (1999). *The Understanding by Design Handbook.* Alexandria, VA.: Association for Supervision and Curriculum Development.

Zawojewski, J.S., and Shaughnessy, J.M. (2000). Mean and median: Are they really so easy? *Mathematics Teaching in the Middle School*, 5, 436–440.

GLOSSARY

acute angle: an angle that is less than 90°

acute triangle: a triangle with three acute angles

addend: one of the numbers added in an addition equation; e.g., in 3 + 4 = 5, both 3 and 4 are addends

algorithm: a systematic procedure or set of steps, usually for carrying out a computation; there are multiple algorithms for any operation

angle: a geometric figure made up of two rays or line segments (called arms) that have the same end point (called the vertex); the size of an angle is a measure of how much one arm is turned away from the other

angle bisector: a ray or line segment that cuts an angle in half to form two congruent angles

area: the number of two-dimensional units that will cover a 2-D shape

area model (probability): a visual system for determining all possible outcomes in a compound event in order to determine theoretical probability; e.g., see **p. 206**

arithmetic sequence: a number pattern in which the difference between consecutive terms is constant; e.g., for 3, 8, 13, 18, 23, ..., the constant difference is 5

arm: See *angle*.

associative property: a property of both addition and multiplication whereby you can change the grouping of the numbers without changing the result; i.e., $(a + b) + c = a + (b + c)$ and $(a \times b) \times c = a \times (b \times c)$

attribute: a characteristic or quality of a pattern or shape, such as colour, size, mass, and number of vertices

axis of rotational symmetry: a line around which a 3-D shape can be turned to reproduce its original orientation before it has completed a full turn; e.g., a cube has 13 axes of rotational symmetry; three are shown on **p. 106**

bar graph: a visual way to display data that uses horizontal or vertical bars to represent data values grouped in categories or discrete numerical intervals; e.g., see **pp. 181 to 183**

base: 1. for a 2-D shape, the side that is measured for calculating its area; any side of a 2-D shape can be the base **2.** 3-D shapes: for a prism, the two congruent, opposite, and parallel polygon faces that are used to name the prism; e.g., a rectangle-based prism; for a pyramid, the single polygon face that is used to name the pyramid; e.g., a square-based pyramid; for a cylinder and cone, the circle face(s) **3.** for a power, the number that is multiplied by itself; e.g., the base in 5^3 ($5 \times 5 \times 5$) is 5

base plan: a 2-D bird's-eye view of a structure that shows its footprint and varying heights; e.g., see **p. 110**

bias: when the results of data collection have been influenced by how the data has been collected from a sample so that the results do not reflect the population that the sample is supposed to represent; e.g., collecting data about the favourite activity of elementary school students by just asking girls would have biased results

box and whisker plot: a visual way to display data that shows how the data values in a set of data are spread or distributed by dividing the set of data into four equal groups called quartiles; also called a box plot; e.g.,

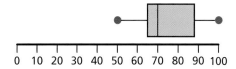

capacity: a measure of the amount of liquid that a container could potentially hold; often measured in millilitres (mL) and litres (L); capacity units are also used to measure amounts of liquids

Carroll diagram: a chart for sorting that allows for cross-classification; e.g., see **p. 169**

centre of rotation: See *turn centre*.

circle graph: a visual way to display data that shows how the parts of a set of data sorted into categories relate to the whole set of data; sometimes called a pie chart; e.g., see **pp. 187 to 189**

circumference: the name of the boundary or perimeter of a circle; also the length of that boundary

classify: to sort a set of items or data into groups and then name or identity the groups to differentiate them

common denominator: the denominator that is the same for two or more fractions; e.g., $\frac{5}{12}$ and $\frac{7}{12}$ have a common denominator of 12

common factor: a whole number that divides into two or more given numbers with no remainder; e.g., 4 is a common factor of 12 and 4

common multiple: a number that is a multiple of two or more given whole numbers; e.g., 12, 24, 36, ... are common multiples of 4 and 6

common numerator: the numerator that is the same for two or more fractions; e.g., $\frac{5}{17}$ and $\frac{5}{12}$ have a common numerator of 5

commutative property: a property of addition and multiplication whereby you can change the order of the numbers without changing the result; i.e., $a + b = b + a$ and $a \times b = b \times a$

composite number: a number that has more than two factors; e.g., 8 is composite because its factors are 1, 2, 4, and 8. See *prime number*.

composite shape: a shape that appears to be composed of multiple simple shapes such as triangles, squares, circles and partial circles

compound event: an event that consists of two or more simple events; e.g., the event of rolling both a 4 and a 6 in two rolls of a die is a compound event. See *dependent events* and *independent events*.

concave: See *convex*.

cone: a 3-D shape that consists of one circle base, one curved edge, one lateral curved surface, and a point called the apex

congruent: geometric figures that are equivalent; 2-D shapes or 3-D shapes are congruent if they have the same shape and size, angles are congruent if they have the same angle measure, and line segments (e.g., sides or edges) are congruent if they are the same length

conjecture: a statement that has not been proved to be true nor shown to be false; e.g., any even number greater than 2 is the sum of two prime numbers

construct: formally, this means to draw using only straight edge (not a ruler) and compass and sometimes a Mira or geometry software; informally, this means to draw or build

convex: a polygon with all interior angles less than 180°; a polygon that is not convex is concave; e.g.,

concave   convex

coordinate grid: a method of locating points on a plane using two numbers; a point can be located by its distance from a horizontal line and a vertical line, each called an axis; e.g., the grid below shows two points plotted on one of the four quadrants in the coordinate grid system

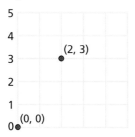

core of a pattern: the part of the pattern that repeats in a repeating pattern; e.g., in the first pattern below, the core is blue square–yellow circle; in the second pattern, there is a colour core: blue-blue-yellow, and a shape core: circle-square

curved edge: where the curved surface of a 3-D shape meets a face; e.g., a cone has one curved edge, and a cylinder has two curved edges (see *cylinder*)

curved surface: a surface of a 3-D shape that is not flat; a cone and cylinder each have one curved surface; e.g., see *cylinder*

cylinder: a 3-D shape that has two congruent, opposite, and parallel circle faces (bases), two curved edges, and one lateral curved surface; e.g.,

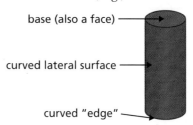

base (also a face)

curved lateral surface

curved "edge"

data spread: how the data values in a set of data are distributed over the range of the set of data including how all the data values relate to the median data value; sometimes called data distribution

data value: a piece of data in a set of data; e.g., in the set of data 3, 6, 9, 15, one data value is 9

denominator: the number below the bar in a fraction; it tells the number of equal parts in the whole; e.g., the denominator of $\frac{3}{4}$ is 4

dependent variable: when graphing two sets of related data, the dependent variable is the one that is affected by changes in the *independent variable*; the dependent variable is conventionally graphed along the vertical axis; e.g., in a graph of distance over time, distance would be the dependent variable because it changes as time elapses

diagonal (of a polygon): a line segment connecting any two non-adjacent vertices of a polygon

diameter: the name of a line segment that travels through the centre of a circle to connect any two points on the circumference; also the length of that line segment

difference: the result of a subtraction; e.g., in $12 - 7 = 5$, 5 is the difference

digit: one of these ten symbols: 0, 1, 2, 3, 4, 5, 6, 7, 8, 9, used to create numerals, e.g., the digits 3, 4, and 5 in the numeral 345

dilatation: a transformation that reduces or enlarges a shape resulting in an image that is similar; see **p. 126**

displacement: a way to measure the volume, usually of an irregular object; when an object is immersed in water, it displaces an amount of water equivalent to its volume (each 1 cm³ of volume displaces 1 mL of water)

distributive property: the property of distributing one operation over another; $a(b + c) = ab + ac$, $a(b - c) = ab - ac$, $\frac{b + c}{a} = \frac{b}{a} + \frac{c}{a}$, and $\frac{b - c}{a} = \frac{b}{a} - \frac{c}{a}$

dividend: the number you start with in a division; e.g., in $24 \div 8 = 3$, 24 is the dividend

divisibility test: a rule for telling whether a number is divisible by another without actually performing that division; e.g., a number is divisible by 9 if the sum of its digits is divisible by 9

divisible: a whole number that can be divided by another number with no remainder is said to be divisible by that number, it is also a multiple of that number; e.g., since 32 is divisible by 4, then 32 is a multiple of 4

divisor: the number you divide by in a division; e.g., in $24 \div 8 = 3$, 8 is the divisor

double bar graph: a special type of bar graph used to compare multiple sets of data; e.g., see **p. 183**

edge: a line segment on a 3-D shape where two faces meet; e.g., a triangle-based prism has nine edges (see *prism*)

equation: a mathematical sentence that has an equals sign and includes one or more variables, also called an algebraic equation; e.g., $3x \div 5 = 10$ and $3x + 4y = 50$

equilateral triangle: a triangle with three congruent sides and three congruent angles; e.g., see **p. 96**

equivalent fractions: different fractions that represent the same part of the same whole; e.g., $\frac{3}{4}$ and $\frac{6}{8}$

Euclidean transformation: a transformation that does not change the size of a shape such as a translation, reflection, and rotation; a non-Euclidean transformation, like a dilatation, changes the size

event: an outcome or set of outcomes that are part of all possible outcomes in a probability experiment; e.g., if the event is rolling a number less than 4 on a die, the possible outcomes are 1, 2, 3, 4, 5, and 6, and the event consists of the outcomes 1, 2, and 3

expanded form: a way to write a number that shows the value of each digit; e.g., 5467 is $5000 + 400 + 60 + 7$ or 5 thousands + 4 hundreds + 6 tens + 7 ones

experimental probability: a number that describes the fraction of the time a favourable outcome happens in the total number of trials in an experiment; e.g., if an even number is rolled 12 times in 20 rolls, the experimental probability of rolling an even number is $\frac{12}{20}$. See *theoretical probability*.

exponent: an exponent tells how many times the base in a power is used as a factor; e.g., in the power 4^3, the exponent 3 indicates that 4 appears 3 times, $4^3 = 4 \times 4 \times 4$

expression: a mathematical phrase that includes one or more variables, also called an algebraic expression; e.g., $2x + 5$ and $5y - 7x$

extrapolate: make a prediction about data beyond the known data values

face: a flat part of a 3-D shape; e.g., a cone has one face and a triangle-based prism has five faces (see *prism*)

factor: in a multiplication sentence, one of the numbers being multiplied; e.g., in $3 \times 4 = 12$, 3 and 4 are factors; also a number that divides into a whole number with no remainder; e.g., both 3 and 4 are factors of 12

formula: an algebraic equation that generalizes a relationship, often in measurement; e.g., the formula *Volume of a prism = Area of base × height* describes the relationship between the area of the base, the height, and the volume of a prism

geometric property: an attribute that all shapes in a certain classification share; e.g., all quadrilaterals have four sides and four vertices and all rectangles have four sides, four vertices, and four right angles

geometric sequence: a number pattern in which the ratio between consecutive terms is constant; e.g., in the pattern 3, 6, 12, 24, 48, ..., the ratio is 2

hexagon: any six-sided polygon; a regular hexagon has six congruent sides and six congruent angles

histogram: a visual way to display data using bars, usually vertical, to represent data grouped in continuous numerical intervals, often measurement data; e.g., see **p. 184**

image: the result of a transformation; e.g., a reflection image

improper fraction: a fraction that has a numerator that is equal to or greater than the denominator; e.g., $\frac{6}{6}$ and $\frac{8}{5}$

independent variable: when graphing two sets of related data, changes in the independent variable affect the *dependent variable*; the independent variable is conventionally graphed along the horizontal axis; e.g., in a graph of distance over time, time would be the independent variable

inequality: a mathematical sentence that has an inequality sign and includes one or more variables; e.g., $3x \div 5 < 10$ and $3x + 4y = 50$

infer: to analyze a set of data and use reasoning to draw conclusions about the data

integers: the set of counting numbers (positive integers), their opposites (negative integers), and 0; ..., -4, -3, -2, -1, 0, 1, 2, 3, 4, ...

interpolate: estimate data between known data values

interval bar graph: a bar graph that groups data into discrete numerical intervals instead of categories. See *bar graph*.

inverse operations: operations that undo each other; addition and subtraction are inverse operations, so are multiplication and division

isometric drawing: a 2-D representation of a 3-D shape drawn on special dot paper; vertical edges on the 3-D shape are vertical in the drawing, width and depth edges on the 3-D shape are diagonal in the drawing, and equal lengths on the 3-D shape are equal in the drawing; see **p. 109**

isosceles triangle: a triangle with two congruent sides; e.g., see **p. 96**

kite: a quadrilateral with two pairs of congruent sides and no parallel sides; e.g., see **pp. 98 and 99**

lateral face: a face on a prism or a pyramid that is not a base

lateral surface: a curved surface on a 3-D shape, such as a cone or cylinder, that is not a base; e.g., see *cylinder*

line: a set of points that form a straight pattern that extends infinitely in both directions

line graph: a visual way to display data that uses a broken line to show relationship and trends, usually over time; e.g., see **p. 189**

line of symmetry: a line that divides a 2-D shape in half so that, if you were to fold along the line, the halves would match; some shapes have multiple lines of symmetry

line segment: a part of a line that has two endpoints; e.g., the sides of a polygon are line segments and the edges of a polyhedron are line segments

lowest common multiple (LCM): the least number that is a multiple of two or more given whole numbers; e.g., 12 is the LCM of 4 and 6

mass: the amount of matter in an object, often measured in grams and kilograms; the mass of an object never changes as opposed to weight, which is influenced by gravity

mean: a measure of central tendency; a single data value used to represent a set of data; one way to determine the mean is to share the total of the data values among the number of data values; e.g., the mean of 3, 7, 12, 14 is 9 because $(3 + 7 + 12 + 14) \div 4 = 9$

measure of central tendency: mean, median, and mode are all measures of central tendency used to represent a set of data

median: a measure of central tendency; a single data value used to represent a set of data; it is the middle value when the data values are ordered from least to greatest; e.g., the mean of 3, 4, 8, 12, 15 is 8

minuend: the number that you start with in a subtraction; e.g., in $15 - 12 = 3$, 15 is the minuend

mixed number: a number that consists of a whole number and a fraction; e.g., $5\frac{1}{2}$

mode: a measure of central tendency; the data value or values that occur most often in a set of data; e.g., the mode in 3, 3, 3, 6, 8, 9 is 3; the modes in 4, 4, 5, 5, 7, 9, 10 are 4 and 5, and there is no mode in 3, 7, 9, 10

multiple: the product of a number and a whole number; e.g., 12 is a multiple of 4 because $3 \times 4 = 12$

net: a model of the 2-D surface of a 3-D shape; if you assemble a net for a shape, it will form the 3-D shape; e.g., see **pp. 107 and 108**

numeral: a symbol that names a number; e.g., the symbol 15 is the numeral that names the number 15

numerator: the number above the bar in a fraction, it tells the number of equal parts in the whole that the fraction describes; e.g., the numerator of $\frac{3}{4}$ is 3

obtuse angle: an angle between 90° and 180°

obtuse triangle: a triangle with one obtuse angle

ordered pair: a pair of numbers that describes the location of a point in a coordinate grid; sometimes called coordinates; e.g., the ordered pair (2, 3) describes a location that is 2 units horizontally to the right of the origin and 3 units vertically above the origin (see **p. 119**)

orientation: when the orientation of a shape changes, its vertices are in a different order; e.g., you can tell the reflection image below has a different orientation than the original shape because when you read the vertices in a clockwise direction starting at A or its image A', the original shape reads ABCD and the image reads A'D'C'B'

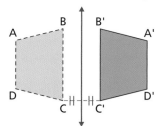

original shape   reflection image

origin: the intersection of the horizontal and vertical axes in a coordinate grid represented by the ordered pair (0, 0); e.g., see **p. 119**

orthographic drawings: a set of drawings that describe a 3-D shape by showing different views of the shape, usually from the top, sides, and front; e.g., see **p. 109**

outcome: a result of an experiment; e.g., when you roll a die, there are six possible and equal outcomes: 1, 2, 3, 4, 5, and 6

parallel: lines and planes that will never meet when extended are parallel; e.g., the opposite sides of a rectangle are parallel line segments and the opposite faces of a cube are parallel planes

parallelogram: a quadrilateral that has two pairs of opposite congruent parallel sides; e.g., see **pp. 98 and 99**

pattern rule: a description of a pattern that results in a unique pattern, it tells how the pattern starts and how it continues; e.g., the pattern rule for 2, 4, 8, 16, ... is start at 2 and multiply by 2 each time

percent: a special ratio that compares a number to 100; e.g., 9 out of 10 is 90 out of 100 or 90%

perimeter: the name of the boundary of a shape; it is also the length of that boundary; circumference is a special name for the perimeter of a circle

perpendicular bisector: a line, ray, or line segment that divides a line segment into two congruent line segments and is perpendicular (at 90°) to that line segment

pictograph: a visual way to display data that uses the same picture or symbol throughout the graph; it represents a set of data that is grouped in categories; see **p. 180**

place value: the placement of a digit in a numeral that tells its value in the numeral; e.g., the place value of 5 in 3<u>5</u>8 is tens so 5 has a value of 5 tens or 50

plane: a flat surface with no depth that extends infinitely in all directions; e.g., the six faces of a cube are each located on a different plane (see **p. 100**)

plane symmetry: a 3-D shape is said to have plane symmetry if there is at least one plane about which the shape is symmetrical; these planes are called planes of symmetry; see **p. 104**

polygon: a closed 2-D shape with only straight sides; e.g., triangles, quadrilaterals, and pentagons are all types of polygons

polyhedron: a 3-D shape that has all polygon faces; e.g., prisms and pyramids are polyhedrons but cones and cylinders are not (see **p. 94**)

power: a number (called a base) raised to an exponent; e.g., 3^4 is a power of 3 and means $3 \times 3 \times 3 \times 3$

power of 10: a power that has a base of 10; e.g., 1000 is a power of 10 because it can be written as 10^3

precision: the smallest unit used in a measurement; e.g., if a measurement is reported as 3.2 cm, the precision is to 0.1 cm, or one tenth of a centimetre, and you can assume that it could have been rounded from a measurement that is anywhere from 3.15 to 3.25

prime number: a whole number that has exactly two factors; e.g., 2 is a prime number because 2 can only be written as a product of 1 and 2; the number 1 is not a prime number, it is called a unit. See *composite number*.

prism: a 3-D shape with two opposite congruent parallel polygon bases and parallelograms (usually rectangles) for lateral faces; e.g.,

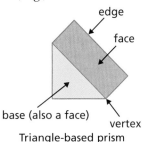

Triangle-based prism

product: the result of a multiplication; e.g., in $3 \times 4 = 12$, 12 is the product

proper fraction: a fraction that has a numerator that is less than the denominator; e.g., $\frac{1}{8}$ and $\frac{8}{9}$

proportion (solve): a mathematical statement that equates two ratios, $\frac{a}{b} = \frac{c}{d}$; to solve a proportion means to determine one of the unknowns in the proportion; e.g., if $\frac{x}{20} = \frac{16}{80}$, then $x = 4$

pyramid: a 3-D shape with one polygon base and triangles for lateral faces

Pythagorean theorem: a statement that relates the lengths of the three sides of a right triangle: the sum of the square of the two legs is equal to the square of the hypotenuse, $a^2 + b^2 = c^2$ (see **p. 116**)

quadrilateral: a four-sided polygon; e.g., squares, trapezoids, and kites are quadrilaterals (see **pp. 98 and 99**)

quartile: one of four equal groups that a set of data can be divided into to be able to see how the data is spread; quartiles are used to construct box plots; e.g., see **pp. 177 and 178**

quotient: the result of a division; e.g., in $20 \div 5 = 4$, 4 is the quotient

radius: the name of the line segment that connects the centre of a circle to any point on the circumference, also the length of that segment

random: anything that happens by chance and is not predictable with certainty; e.g., the result of rolling a die

range: the difference between the greatest and least data values in a set of data; e.g., in the set of data 3, 7, 19, 20, 45, the range is 42 $(45 - 3 = 42)$

rate: a comparison or ratio of two values, each with a different kind of unit; e.g., \$2 per dozen and 50 km/h

ratio: a comparison of two quantities; e.g., if there are 20 girls and 25 boys in a group, the ratio of girls to boys is 20 to 25 or 4:5

ray: part of a line with one end point that extends infinitely at the other end; e.g.,

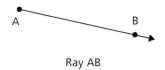

Ray AB

reciprocal: the reciprocal of the fraction $\frac{a}{b}$ is $\frac{b}{a}$

rectangle: a special parallelogram with four right angles; e.g., see **pp. 98 and 99**

rectangle-based prism: a prism that has two opposite, congruent, and parallel rectangle bases

recursive pattern: a pattern where each term in the pattern is defined by a previous term; e.g., the pattern defined by the rule: start at 2 and add 2 each time (2, 4, 6, 8,...) is a recursive pattern

reflection: a transformation that flips a shape across a reflection line without turning or sliding and results in a congruent mirror image with the opposite orientation; informally called a flip; e.g., see **pp. 122 to 124**

reflection line: the line across which a shape is reflected or flipped. See *reflection*.

reflex angle: an angle that is greater than 180° but less than 360°

regular polygon: a polygon that has congruent sides and congruent angles

repeating decimal: a decimal that repeats and never terminates; e.g., 0.232323... and 0.135613561356...

rhombus: a parallelogram with four congruent sides; e.g., see **pp. 98 and 99**

right angle: a 90° angle

right triangle: a triangle with one 90° angle

rotation: a transformation that turns a shape around a turn centre and results in an image that is congruent to the original shape with the same orientation; informally called a turn; e.g., see **pp. 124 to 126**

sample size: the size of a sample that is selected to represent a population in a data collection situation; the larger the sample size, the more likely one will be able to generalize the results to the full population it represents

sample space: the set of all possible outcomes in an experiment; e.g., the sample space of flipping two coins has four outcomes: HH, HT, TH, and TT

scale: on a bar graph, an arrangement of numbers in regular intervals that tells the value of the lengths of the bars; on a pictograph, the scale is the value of each symbol

scalene triangle: a triangle with three different side lengths; e.g., see **p. 96**

scatter plot: a visual way to display data to see if there is a relationship between two sets of data; the pattern of the plotted points will tell if there is a relationship and what kind of relationship it might be; e.g., see **p. 190**

scientific notation: a number expressed as the product of a multiplier that is from 1 to less than 10 and a power of 10; e.g., 1245 in scientific notation is 1.245×10^3 and 0.00342 in scientific notation is 3.42×10^{-3}

side: a part of a polygon; e.g., a triangle has three sides

similar: shapes that are the same shape but not necessarily the same size are similar; if two polygons are similar, each pair of corresponding side lengths has the same ratio

simplest terms: also called simplest form or lowest terms **1.** when the numerator and denominator in a fraction have no common factors other than 1; e.g., $\frac{2}{3}$ is in simplest terms but $\frac{5}{10}$ is not **2.** when the terms in a ratio have no common factors other than 1; e.g., 3:4 is in simplest terms

simulation: an experiment used to estimate the probability of a situation by representing events with other simpler ones of equal probability; e.g., see **pp. 206 to 208**

skeleton: a model of a 2-D shape or 3-D shape that emphasizes certain components; in a 2-D skeleton, the vertices and sides are emphasized, and in a 3-D skeleton, the vertices and edges are emphasized; e.g., see **pp. 106 to 107**

solve an equation: to determine the value of an unknown in an equation; e.g., to determine the value of x in $4 + 2x = 10$

sphere: a 3-D shape in the form of a ball; mathematically it is defined as the set of points in space that are a given distance from a fixed point, the centre; the terms *radius* and *diameter* also apply to a sphere

square: a rectangle with four congruent sides; also a rhombus with four right angles; e.g., see **pp. 98 and 99**

standard form: the usual form of a number; e.g., 3456 and 0.0467 are in standard form. See *expanded form* and *scientific notation.*

standard unit: a measurement unit that is common to everyone; e.g., a measurement such as 3 m is a standard measure where a measure such as 3 paces is not

statistic: at the elementary school level, students work with statistics such as mean, median, mode, and range to represent and analyze sets of data

stem-and-leaf plot: a visual way to display data that groups the data in place value intervals and shows each data value in the set; e.g., the stem and leaf plot below graphs a set of 29 data values about test scores that range from a score of 101 to a score of 128

10	1 2 2 7 8 9
11	0 3 4 4 5 8 9 9 9
12	0 0 0 0 1 1 2 2 3 3 3 5 5 8

straight angle: an angle that is exactly 180°

subtrahend: in a subtraction, the number that is subtracted; e.g., in $12 - 4 = 8$, 4 is the subtrahend

sum: the result of an addition; e.g., in $2 + 3 = 5$, the sum is 5

surface area: the total area on the outside of a 3-D shape; it is a 2-D measure, often in cm^2 and m^2; e.g., the net below represents the surface area of a pyramid

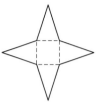

symmetry: 1. reflective symmetry can be 2-D and 3-D, also called line symmetry (2-D), plane symmetry (3-D), and mirror symmetry (2-D and 3-D); a 2-D shape has reflective symmetry when it can be folded in half along a line (line of symmetry) and the halves match exactly, a 3-D shape has reflective symmetry when the two halves on either side of a plane (plane of symmetry) are mirror images of each other; e.g., see **pp. 102 to 104. 2.** rotational symmetry can be 2-D and 3-D, also called turn symmetry; a shape has rotational symmetry if it can be turned around its centre point (2-D shape) or a line (axis of symmetry in a 3-D shape) and fit into an outline of itself at least once before it has completed a full turn; e.g., see **pp. 104 to 106**

table of values (T-chart): a way to display numerical data in chart form to emphasize the relationship between two variables

term: 1. each number in a number pattern is called a term **2.** each part of a ratio or rate is a term; e.g., in the ratio 3:4, 3 and 4 are terms **3.** In an expression, each part is a term; e.g., in $3x + 4 = 12$, there are three terms

terminating decimal: a decimal that terminates; e.g., the decimal for $\frac{1}{8}$ is a terminating decimal, 0.125

tessellation: an arrangement of congruent 2-D shapes that covers an area without gaps or overlap; sometimes called a tiling

theoretical probability: a fraction that compares the number of favourable outcomes to the total number of possible outcomes; e.g., the theoretical probability of rolling an even number on a die is $\frac{3}{6}$ or $\frac{1}{2}$ since there are three favourable outcomes (2, 4, 6) and 6 possible outcomes (1, 2, 3, 4, 5, 6). See *experimental probability.*

transformation: a motion that relocates a shape resulting in an image. See *dilatation, reflection, rotation,* and *translation.*

translate a pattern: to change a pattern to an equivalent but different pattern with the same underlying mathematical structure; e.g., the number pattern 1, 2, 2, 1, 2, 2, ... is a translation of the shape pattern square-triangle-triangle, square-triangle-triangle, square-triangle-triangle, ... because both follow an ABB pattern

translation: a transformation that slides a shape without turning or flipping and results in an image that is congruent to the original shape with the same orientation; informally called a slide. See *translation rule.*

translation rule: a description of a translation that results in a unique translation; e.g., the diagonal translation below is described with a translation rule that has two parts (represented by the two dashed red arrows), even though the translation is a single diagonal motion (represented by the single solid red arrow)

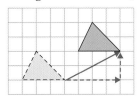

Translation rule: 4 units right, 2 units up

trapezoid: a special quadrilateral that has exactly two parallel sides; an isosceles trapezoid also has two opposite congruent sides; e.g., see **pp. 98 and 99**

tree diagram: a visual method for systematically determining all possible outcomes in a compound event in order to calculate theoretical probability; e.g., see **p. 205**

triangle-based prism: See *prism.*

triangular numbers: numbers in the pattern 1, 3, 6, 10, 15, ...; each term is called a triangular number because that number of items can be formed into a triangle

turn centre: The point at which a shape is turned in a rotation; it is also the point around which a shape is turned to determine rotational symmetry; also called centre of rotation; e.g., see **p. 125**

unit fraction: a fraction with a numerator of 1; $\frac{1}{2}$, $\frac{1}{3}$, $\frac{1}{4}$, ... are the unit fractions

unit rate: a rate with a second term of 1; e.g., the rate 6 km in 3 min can be changed to the equivalent unit rate 2 km/min

unit ratio: a ratio that has one term of 1; e.g., the ratio 2:4 can be changed to the equivalent unit ratio 1:2

variable: a letter or symbol that represents one or more values in an equation or expression; e.g., in $3x + 2 = 20$, the variable x represents 6, in $3 \blacklozenge + 2 > 20$, the variable $\blacklozenge$ represents 7 or more

Venn diagram: a visual system of sorting circles that allows for cross-classification; e.g., see **p. 168**

vertex: **1.** in a polygon, the point where two sides meet **2.** in a 3-D shape, the point where three or more edges meet **3.** in an angle, the point where two arms meet

volume: a measure of the amount of three-dimensional space that an object takes up, often measured in cm³ and m³

zero principle: the principle whereby a pair of opposite integers have a value of zero; this principle is used in many integer operation situations; e.g., see **p. 86**

Index

observation, 171
primary (first-hand) data, 172
quartiles, 177–178
questionnaires, 171
range, 177
sample size, 172–173
secondary (second-hand) data, 172
sorting and classifying, 167–169
sources, first- and second-hand, 172
surveys, 169–172
Decagon, 95
Deca-. *See* Metric system, prefixes
Decimals, 61–73
    addition, 68–71
    base ten place value system, 62–63
    comparing, 67
    creating in-between, 67
    division, 71–73
    equivalent, 64–65
    estimation, 66
    fractions, relating, 65–66
    mental math, 68–73
    multiplication, 71–73
    patterns, 6
    principles, 62–64
    reading, 66
    rounding, 66
    subtraction, 68–71
    writing, 67
Dependent variables, 190–191
Diameter, 138–139
Dilatations, 126–127
Discount, 83
Displacement, 157–158
Distributive property, 29–30, 89–91
Division
    decimals, 71, 73
    fractions, 56, 58–60
    integers, 91–92
    mixed numbers, 60
    powers of ten, 31, 33
    principles, whole numbers, 25–31
    remainders, 40–41
    varied approaches/algorithms, 38–40
    zero, as dividend or divisor, 30–31
Dodecagon, 95

E

Edges, 94
Equations and graphs, 13–14
Equilateral triangles, 96, 100
Equivalent fractions, 46–47, 58, 65–66
Estimation
    area, 141, 143
    decimals, 66

measurement, 131, 141, 143
numbers, 19
products and quotients, 33–34
rounding, 34
strategies, 34
Events, probability, 198–200
Exponents, 24
Extrapolation, 194

F

Faces, 94
Factors, 20–24
    common, 24
    determining, 21
    divisibility, 22
    greatest common, 24
    prime numbers, 23
Fibonacci sequence, 3
Fraction strips, modelling, 52–53, 56, 59, 60
Fractions, 42–60
    addition, 50–55
    decimals, relating, 65–66
    division, 58–60
    equivalent, 46–47, 58, 65–66
    improper, 47
    meanings, 43
    multiplication, 56–58
    patterns, 6
    principles, 44–46
    principles for comparing, 48–50
    probability, 199–200
    subtraction, 54–55
Functions, 9–10

G

Geoboards, 142, 149
Geometric sequences, 2
Goldbach's conjecture, 23
Gradients, angle measure, 163
Graphs, 13–14, 179–192
    conclusions from, 192–194
    equations, 13–14
    extrapolation and interpolation, 194
    misleading, 194–196
    rates and ratios, 76
    relationships, 13–14
Greatest common factor (GCF), 24

H

Hendecagon, 95
Heptagon, 95
Hexagon, 95, 100
    regular, 100
Histograms, 179, 184–185

I

Improper fractions, 47
Independent variables, 190–191
Inequalities, solving, 13
Integers, 84–92
    addition, 86–87
    comparing, 85–86
    contexts, 84
    division, 91–92
    multiplication, 89–91
    principles for comparing, 85–86
    reading and writing, 85
    subtraction, 88–89
    zero principle, 86, 88
Interpolation, 194
Inverse operations, 27
Isometric drawings, 106, 109
Isosceles triangles, 96

K

Kilo-. *See* Metric system, prefixes
Kilogram, SI Base unit, 134
Kites, 98, 99

L

Lateral surface, 94
Length, 136–139
Line graphs, 179, 189
Line of symmetry, 102
Lines and line segments, 100–101
    bisector, 117
    intersecting, 101
    parallel, 101, 117
    perpendicular, 101, 117
    symmetry, 101–106
Litres, 134, 152
Location, 118–120
Lowest common multiple (LCM), 24

M

Maps, 118–119
Mass, measurement of, 158
Mean, 173–175
    See also Average; Median; Mode
Measurement, 129–166
    angles, 159–166
    area, 139–150
    benchmarks, 129, 131
    capacity, 152
    definition/comparison stage, 132–133
    estimation, 129, 131
    length, 136–139
    mass, 158

    metric system, 133–135
    non-standard units stage, 132, 133
    perimeter, 137–139
    precision, 137
    stages, 132–133
    standard units stage, 132, 133
    volume, 153–158
Measures of central tendency, 173–176
Measures of data spread, 177–178
Median, 176
    See also Average; Mean; Mode
Mental calculation
    decimals, 73
    whole numbers, 41
Metric system
    measurement, 133–135
    prefixes, 135
    units, 134
Milli-. *See* Metric system, prefixes
Millilitres, 152, 158
Mixed numbers, 47, 54–55
    addition and subtraction, 54–55
    division, 60
    multiplication, 58
Mode, 176
    See also Average; Mean; Median
Multiples, 20, 24
Multiplication
    decimals, 71–73
    fractions, 56–58
    integers, 89–91
    lattice algorithm, 37–38
    mixed numbers, 58
    powers of ten, whole numbers, 31–32
    principles, whole numbers, 26–31
    varied approaches/algorithms, 34–38
    zero as a factor, 30

N

Nets, 107–108
Nonagon, 95
Number lines
    comparing integers, 85
    fraction operations, 51, 54–55
    integer operations, 86, 88–92
Number theory, 20–24
Numeration principles, whole numbers, 15–18

O

Obtuse angles, 166
Obtuse triangles, 96
Octagon, 95
Ordered pairs, 119–120
Orthographic drawings, 109
Outcomes, probability, 198–199

Scale, 180, 181
    misuse of, 195–196
Scalene triangles, 96
Scatter plots, 179, 190–191
Scientific notation, 17
Shapes
    attributes of, 93
    combining and dissecting, 110–111
    properties of, 93–100
    representing, 106–108
SI (Système International). *See* Measurement;
    Metric system
Sieve of Eratosthenes, primes, 23
Similarity, 112, 114–116
Simulation experiments, 206–208
Skeletons, 2-D and 3-D, 106–107
Spheres, 94
    volume, 156–157
Squares, 99, 100
    area, 147
Standard form, 17–18
Statistics, 173–178
Stem-and-leaf plots, 179, 185–186
Straight angles, 166
Subtraction
    decimals, 69–71
    fractions, 50–54
    integers, 88–89
    mixed numbers, 54–55
Surface area, 150–151
    prisms, 150–151
    pyramids, 150
    volume, relating, 157
Surveys, 169–171
Symmetry, 101–106
    line, 101–103, 105
    plane, 104
    reflective, 101–104, 105
    rotational, 101, 104–106

T

Tangrams, 111
Tessellations, 127–128
    transformations, relating, 128
Transformations, 120–127
    coordinate grids, 121, 122
    tessellations, relating, 128
Translations, 120–122
Transparent mirrors, 124

Trapezoids, 98, 99
    area, 149
Tree diagrams, 205
Triangles, 95–98
    acute, 96
    classifying, 96
    equiangular, 96
    equilateral, 96, 100
    isosceles, 96
    obtuse, 96
    right, 96
    rigidity, 98
    scalene, 96
    sum of angles, 97
Triangular numbers, 3

U

Unit rates, 78

V

Variables
    algebra, 7–8
    dependent and independent, 190
Venn diagrams, 98, 168
Vertices, 94, 95
Volume
    capacity, relating, 153, 156, 157–158
    cones, 156–157
    cylinder, 155–157
    formulas, 155–157
    non-standard units, 153
    prisms, 155–157
    pyramids, 156–157
    spheres, 157
    standard units, 154–155
    surface area, relating, 157

W

Weight, versus mass, 158
Whole number operations, 25–41

Z

Zero
    division, 30–31
    multiplication, 30
Zero principle, integers, 86